Machine Learning for Cloud Management

Machine Learning for Cloud Management

Jitendra Kumar
Ashutosh Kumar Singh
Anand Mohan
Rajkumar Buyya

CRC Press
Taylor & Francis Group
Boca Raton London New York

CRC Press is an imprint of the
Taylor & Francis Group, an **informa** business

A CHAPMAN & HALL BOOK

First edition published 2022
by CRC Press
6000 Broken Sound Parkway NW, Suite 300, Boca Raton, FL 33487-2742

and by CRC Press
2 Park Square, Milton Park, Abingdon, Oxon, OX14 4RN

Library of Congress Cataloging-in-Publication Data

Names: Kumar, Jitendra, 1975- author. | Singh, Ashutosh Kumar, author. |
Mohan, Anand (Of Indian Institute of Technology), author. | Buyya,
Rajkumar, 1970- author.
Title: Machine learning for cloud management / Jitendra Kumar, Ashutosh
Kumar Singh, Anand Mohan, Rajkumar Buyya.
Description: First edition. | Boca Raton : CRC Press, 2022. | Includes
bibliographical references and index.
Identifiers: LCCN 2021027713 | ISBN 9780367626488 (hardback) | ISBN
9780367622565 (paperback) | ISBN 9781003110101 (ebook)
Subjects: LCSH: Cloud computing. | Machine learning.
Classification: LCC QA76.585 .K85 2022 | DDC 004.67/82--dc23
LC record available at https://lccn.loc.gov/2021027713

ISBN: 978-0-367-62648-8 (hbk)
ISBN: 978-0-367-62256-5 (pbk)
ISBN: 978-1-003-11010-1 (ebk)

DOI: 10.1201/9781003110101

Dedicated to,
My wife: Gita, daughter: Aru, and Parents

~Jitendra Kumar

Anushka, Aakash, Akankshya, and Parents

~Ashutosh Kumar Singh

My wife: Sudha Mohan, son: Ashish Mohan, daughter: Amrita Mohan, and Late parents

~Anand Mohan

My international collaborators and team members in Melbourne CLOUDS Lab!

~Rajkumar Buyya

Contents

List of Figures

List of Tables

Preface

Cloud computing has become one of the revolutionary technology in the history of the computing world. It offers subscription-based on-demand services and has emerged as the backbone of the computing industry. It has enabled us to share resources among multiple users through virtualization by the means of creating a virtual instance of a computer system running in an abstracted hardware layer. Unlike early distributed computing models, it assures limitless computing resources through its large-scale cloud data centers. It has gained wide popularity over the past few years, with an ever-increasing infrastructure, number of users, and amount of hosted data. The large and complex workloads hosted on these data centers introduce several challenges: resource utilization, power consumption, scalability, operational cost, and many others. Therefore, a practical resource management scheme is essential to bring operational efficiency with improved elasticity. The elasticity of a system depends on several factors such as the accuracy of anticipated workload information, performance behavior of applications in different scenarios communicating the forecast results, use of the anticipated information, and many others.

Effective resource management can be achieved through workload prediction, resource scheduling, and provisioning, virtual machine placement, or a combination of these approaches. The workload prediction has been widely explored and a number of methods are presented. However, the existing methods suffer from various issues including the incapability of capturing the non-linearity of workloads and iterative training that consumes huge computing resources and time. This book discusses the machine learning-based approaches to address the above-mentioned issues. The highlights of the discussed models are continuous learning from error feedback, adaptive nature, decomposition of workload traces, and ensemble learning. Detailed analysis of predictive methods on different workload traces is also included and their performance is compared with state-of-art models. Virtual machine placement is another aspect that is explored to achieve efficient resource management. In general, virtual machine placement is a multiobjective problem that involves multiple conflicting objectives to be optimized simultaneously. The frameworks discussed in this book address the issues of resource utilization, power consumption, and security while placing the workloads on servers.

The remainder of the book is organized as follows: Chapter 1 briefs the basic cloud computing concepts. The discussion on the workload prediction mechanisms begins in chapter 2. First, the basic time series forecasting models are discussed with their performance on different workload traces. Chapter 3 discusses the error preventive time series forecasting models which significantly improve the performance over classical time series models. Then, a discussion on various nature-inspired algorithms is included in chapter 4. It also evaluates the performance of neural network-based forecasting models trained by these algorithms. Next, the forecasting models trained by adaptive differential evolution are presented in Chapter 5. The first learning algorithm allows learning the best suitable mutation strategy and crossover rate. In contrast, the second algorithm allows learning both crossover and mutation strategies along with mutation and crossover rates. Chapter 6 discusses the blackhole neural network-based forecasting scheme and evaluates its performance on several workload traces. It also discusses the concept of self-directed learning. Also, it discusses the self-directed workload forecasting model inspired by an error preventive scheme along with a modification in the blackhole learning algorithm to improve the learning capability of the model. Chapter 7 introduces the decomposition and ensemble learning-based models. The decomposition-based model trains one network for each component extracted from the decomposition of workload trace whereas the second approach creates an ensemble of extreme learning machines and weights their opinions using a blackhole learning algorithm. Chapter 8 introduces two multi-objective load balancing frameworks. The first framework considers the resource utilization and power consumption as objectives to be optimized whereas the second framework also considers the security aspect into consideration while assigning the VMs to servers. The framework deals with side-channel attacks only and minimizes the likelihood of the attack occurring. It also ensures to reduce the number of victim users if any attack occurs. Finally, Chapter 9 summarizes the work discussed in the book.

Author

Dr. Jitendra Kumar is an assistant professor in machine learning at the National Institute of Technology Tiruchirappalli, Tamilnadu, India. He obtained his doctorate in 2019 from the National Institute of Technology Kurukshetra, Haryana, India. He is also a recipient of the Director's medal for the first rank in the University examination at Dayalbagh Educational Institute, Agra, Uttar Pradesh in 2011. He has experience of three years in academia. He has published several research papers in international journals and conferences of high repute, including IEEE Transactions on Parallel and Distributed Systems, Information Sciences, Future Generation Computer Systems, Neurocomputing, Soft Computing, Cluster Computing, IEEE-FUZZ, etc. He has also obtained the best paper awards in two international conferences. His research interests are machine learning, cloud computing, healthcare, parallel algorithms, and optimization. He is also a review board member of several journals, including IEEE Transactions on Computers, IEEE Transactions on Parallel and Distributed Systems, IEEE Access, Journal and Parallel Distributed Computing, and more.

Prof. Ashutosh Kumar Singh is an esteemed researcher and academician in the domain of Electrical and Computer engineering. Currently, he is working as a Professor; Department of Computer Applications; National Institute of Technology; Kurukshetra, India. He has more than 20 years of research, teaching, and administrative experience in various University systems of the India, UK, Australia, and Malaysia. Dr. Singh obtained his Ph.D. degree in Electronics Engineering from Indian Institute of Technology-BHU, India; Post Doc from Department of Computer Science, University of Bristol, UK and Charted Engineer from UK. He is the recipient of the Japan Society for the Promotion of Science (JSPS) fellowship for a visit to the University of Tokyo and other universities of Japan. His research area includes Verification, Synthesis, Design, and Testing of Digital Circuits, Predictive Data Analytics, Data Security in Cloud, Web Technology. He has more than 250 publications till now which includes peer-reviewed journals, books, conferences, book chapters, and news magazines in these areas. He has co-authored eight books including "Web Spam Detection Application using Neural Network", "Digital Systems Fundamentals" and "Computer System Organization & Architecture". Prof. Singh has worked as principal investigator/investigator for six sponsored research projects and was a key member on a project from EPSRC (United Kingdom) entitled "Logic Verification and Synthesis in New Framework".

Dr. Singh has visited several countries including Australia, United Kingdom, South Korea, China, Thailand, Indonesia, Japan, and the USA for collaborative research work, invited talks, and present his research work. He had been entitled to 15 awards such as Merit Awards-2003 (Institute of Engineers), Best Poster Presenter-99 in 86th

Indian Science Congress held in Chennai, INDIA, Best Paper Presenter of NSC'99 INDIA and Bintulu Development Authority Best Postgraduate Research Paper Award for 2010, 2011, 2012.

He has served as an Editorial Board Member of International Journal of Networks and Mobile Technologies, International Journal of Digital Content Technology and its Applications. Also, he has shared his experience as a Guest Editor for Pertanika Journal of Science and Technology, Chairman of C'UTSE International Conference 2011, Conference Chair of series of International Conference on Smart Computing and Communication (ICSCC), and as an editorial board member of UNITAR e-journal. He is involved in reviewing processes in different journals and conferences of repute including IEEE transaction of computer, IET, IEEE conference on ITC, ADCOM, etc.

Prof. Anand Mohan has nearly 44 years of experience in teaching and research and the administration and management of higher educational institutions. He is currently an institute professor in the Department of Electronics Engineering, Indian Institute of Technology (BHU), Varanasi, India. Besides his present academic assignment, Prof. Mohan is a Member of the Executive Council of Banaras Hindu University and Vice-Chairman of the Board of Governors of Indian Institute of Technology (BHU), Varanasi, India. Prof. Mohan served as Director (June 2011-June 2016) of the National Institute of Technology (NIT), Kurukshetra, Haryana, India, and was also Mentor Director of the National Institute of Technology, Srinagar, Uttarakhand, India. For his outstanding contributions in the field of Electronics Engineering, Prof. Mohan was conferred the "Lifetime Achievement Award" (2016) by Kamla Nehru Institute of Technology, Sultanpur, India.

Prof. Rajkumar Buyya is a Redmond Barry Distinguished Professor and Director of the Cloud Computing and Distributed Systems (CLOUDS) Laboratory at the University of Melbourne, Australia. He is also serving as the founding CEO of Manjrasoft Pty Ltd., a spin-off company of the University, commercializing its innovations in Cloud Computing. He served as a Future Fellow of the Australian Research Council during 2012-2016. He serving/served as an Honorary/Visiting Professor for several elite Universities including Imperial College London (UK), University of Birmingham (UK), University of Hyderabad (India), and Tsinghua University (China). He received B.E and M.E in Computer Science and Engineering from Mysore and Bangalore Universities in 1992 and 1995 respectively; and a Doctor of Philosophy (Ph.D.) in Computer Science and Software Engineering from Monash University, Melbourne, Australia in 2002. He was awarded Dharma Ratnakara Memorial Trust Gold Medal in 1992 for his academic excellence at the University of Mysore, India. He received Richard Merwin Award from the IEEE Computer Society (USA) for excellence in academic achievement and professional efforts in 1999. He received Leadership and Service Excellence Awards from the IEEE/ACM International Conference on High-Performance Computing in 2000 and 2003. He received the "Research Excellence Awards" from the University of Melbourne for productive and quality research in computer science and software engineering in 2005 and 2008. With over 112,400 citations, a g-index of 322, and an h-index of 145, he is the highest cited computer scientist in Australia and one of the world's Top 30 cited authors in computer science and software engineering. He received the Chris Wallace Award for Outstanding Research Contribution 2008 from the Computing Research and Education Association of Australasia, CORE, which is an association of university departments of computer science in Australia and New Zealand. Dr. Buyya received the "2009 IEEE TCSC Medal for Excellence in Scalable Computing" for pioneering the economic paradigm for utility-oriented distributed computing platforms such as Grids and Clouds. He served as the founding Editor-in-Chief (EiC) of IEEE Transactions on Cloud Computing (TCC). Dr. Buyya is recognized as a "Web of Science Highly Cited Researcher" for five consecutive years since 2016, a Fellow of IEEE and Scopus Researcher of the Year 2017 with Excellence in Innovative Research Award by Elsevier, and "Lifetime Achievement Award" from two Indian universities for his outstanding contributions to Cloud computing and distributed systems. He has been recently recognized as the "Best of the Worl", in the Computing Systems field, by The Australian 2019 Research Review.

Dr. Buyya has authored/co-authored over 850 publications. Since 2007, he received twelve "Best Paper Awards" from international conferences/journals including a "2009 Outstanding Journal Paper Award" from the IEEE Communications Society, USA. He has co-authored five text books: Microprocessor x86 Programming (BPB Press, New Delhi, India, 1995), Mastering C++ (McGraw Hill Press, India, 1st edition in 1997 and 2nd edition in 2013), Object Oriented Programming with Java: Essentials and Applications (McGraw Hill, India, 2009), Mastering Cloud Computing (Morgan Kaufmann, USA; McGraw Hill, India, 2013; China Machine Press, 2015), and Cloud Data Centers and Cost Modeling (Morgan Kaufmann, USA, 2015). The books on emerging topics that he edited include High Performance Cluster Computing (Prentice Hall, USA, 1999), High Performance Mass Storage and Parallel I/O (IEEE and Wiley Press, USA, 2001), Content Delivery Networks (Springer, Germany, 2008), Market Oriented Grid and Utility Computing (Wiley Press, USA, 2009), and Cloud Computing: Principles and Paradigms (Wiley, USA, 2011). He also edited proceedings of over 25 international conferences published by prestigious organizations, namely the IEEE Computer Society Press (USA) and Springer Verlag (Germany). He served as Associate Editor of Elsevier's Future Generation Computer Systems Journal (2004-2009) and currently serving on editorial boards of many journals including Software: Practice and Experience (Wiley Press). Dr. Buyya served as a speaker in the IEEE Computer Society Chapter Tutorials Program (from 1999-2001), Founding Co-Chair of the IEEE Task Force on Cluster Computing (TFCC) from 1999-2004, and member of the Executive Committee of the IEEE Technical Committee on Parallel Processing (TCPP) from 2003-2011. He served as the first elected Chair of the IEEE Technical Committee on Scalable Computing (TCSC) during 2005-2007 and played a prominent role in the creation and execution of several innovative community programs that propelled TCSC into one of the most successful TCs within the IEEE Computer Society. In recognition of these dedicated services to the computing community over a decade, the President of the IEEE Computer Society presented Dr. Buyya a Distinguished Service Award in 2008.

Dr. Buyya has contributed to the creation of high-performance computing and communication system software for PARAM supercomputers developed by the Centre for Development of Advanced Computing (C-DAC), India. He has pioneered Economic Paradigm for Service-Oriented Distributed Computing and demonstrated its utility through his contribution to conceptualization, design, and development of Grid and Cloud Computing technologies such as Aneka, GridSim, Libra, Nimrod-G, Gridbus, and Cloudbus that power the emerging eScience and eBusiness applications. He has been awarded, over $8 million, competitive research grants from various national and international organizations including the Australian Research Council (ARC), Sun Microsystems, StorageTek, IBM, and Microsoft, CA Australia, Australian Department of Innovation, Industry, Science, and Research (DIISR), and European Council. Dr. Buyya has been remarkably productive in a research sense and has converted much of that knowledge into linkages with industry partners (such as IBM, Sun, and Microsoft), into software tools useful to other researchers in a variety of scientific fields, and into community endeavors. Software technologies for Grid and Cloud computing developed under Dr. Buyya's leadership have gained rapid acceptance and are in use at several academic institutions and commercial enterprises in 50+ countries around the world. In recognition of this, he received Vice Chancellor's inaugural "Knowledge Transfer Excellence (Commendation) Award" from the University of Melbourne in Nov 2007. Manjrasoft's Aneka technology for Cloud Computing developed under Dr.Buyya's leadership has received the "2010 Asia Pacific Frost & Sullivan New Product Innovation Award". Recently, Dr. Buyya received the "Bharath Nirman Award" and the "Mahatma Gandhi Award" along with Gold Medals for his outstanding and extraordinary achievements in Information Technology field and services rendered to promote greater friendship and India-International cooperation.

Abbreviations

SLA	Service Level Aggrement
QoS	Quality of Service
DE	Differential Evolution
PSO	Particle Swarm Optimization
EA	Evolutionary Algorithm
FSA	Firefly Search Algorithm
HS	Harmony Search
TLBO	Teaching Learning Based Optimization
GSA	Gravitational Search Algorithm
BhOA	Blackhole Algorithm
BhNN	Bloackhole Network
MSE	Mean Squared Error
RMSE	Root Mean Squared Error
MAE	Mean Absolute Error
RMAE	Relative Mean Absolute Error
CoC	Correlation Coefficient
SEI	Sum of Elasticity Index
EP	Error Preventive
EPS	Error Prevention Score
NEP	Non Error Preventive
PER	Predictions in Error Range
MoP	Magnitude of Prediction
SaDE	Self Adaptive Differential Evolution
BaDE	Biphase Adaptive Differential Evolution
SDL	Self Directed Learning
ELM	Extreme Learning Machine
RELB	Resource Efficient Load Balancing
SCA	Side Channel Attack
SEALB	Secure and Energy Aware Load Balancing
PWS	Prediction Window Size
WP_{BPNN}	BPNN based Workload Prediction Model
WP_{BhNN}	BhNN based Workload Prediction Model
WP_{BhNN}^{SDL}	BhNN and SDL based Workload Prediction Model
WP_{SaDE}	SaDE based Workload Prediction Model
WP_{BaDE}	BaDE based Workload Prediction Model
ELMNN	ELM based Neural Network

eELMNN	Ensemble of ELM based Neural Networks
WP_{ELMNN}	ELMNN based Workload Prediction Model
WP_{eELMNN}	eELMNN based Workload Prediction Model
$\mathcal{NDS}$	Non Dominated Sorting

CHAPTER 1

Introduction

CLOUD COMPUTING paradigm enables the delivery of computing resources and applications to users across the globe as subscription-oriented services. Virtualization is the technique behind the scene that helps in resource sharing among multiple users in this cloud computing environment.

1.1 CLOUD COMPUTING

Cloud computing is a form of distributed computing environment where multiple virtual instances of a computer system run in abstracted hardware level and every user experiences like owning the entire system. The cloud infrastructure may be private (serves to a single organization), public (shared among multiple organizations), and hybrid (combination of both). A cloud system provides the on-demand services at three different levels, referred to as Infrastructure as a Service (IaaS), Platform as a Service (PaaS), and Software as a Service (SaaS), as shown in Fig. 1.1. In IaaS, the infrastructure components such as servers, networking and storage, and operating services hosted by service providers are delivered to the consumers through virtualization. These components are provided with various services including monitoring, security, log access, backup and recovery, and load balancing. While in PaaS, users get required and associated infrastructure to develop, run, and manage their applications. The service provider is responsible for providing the servers, operating system, storage, database, and middleware such as Java and .NET runtime. In the case of SaaS that is a software distribution model, the software or applications are hosted in the data centers, and users access these applications over the Internet. The applications delivered through SaaS eliminate the requirement of hardware, installation, support, and maintenance as they do not need any installation on local computers and can be accessed through web browsers.

In the last decade, cloud systems have received wide popularity due to ever-growing services, infrastructure, clients, and the ability to host big data [17]. A survey conducted in 2017 reported that organizations would shift their 90% enterprise workload on a cloud by 2021 [29]. The cloud infrastructure is growing very fast, and the cloud industry is expected to grow with 14.6% compound annual growth rate to reach the $300 billion mark by 2022 [62,63]. Modern cloud systems are equipped with characteristics such as on-demand service, reliability, scalability, elasticity,

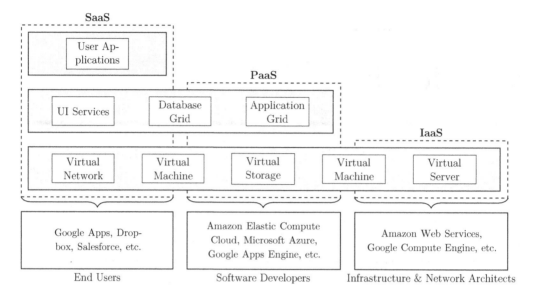

FIGURE 1.1 Service model view of cloud computing

disaster recovery, accessibility, measured services, and many others [25, 73, 75, 98]. However, various challenges and limitations are still open including inefficient resource management, security and privacy, heterogeneity, elasticity, usability, response time, and many more [18, 19, 21, 52--54, 90, 109, 124, 125].

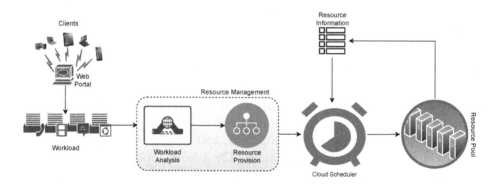

FIGURE 1.2 Cloud resource management view

1.2 CLOUD MANAGEMENT

Resource management is one of the core functions of cloud systems and must be improved for better system performance [66, 102, 119]. The inefficiency in resource management directly affects the system performance and operational cost. The poor resource utilization degrades the overall system performance and may increase the service cost as well. A simple resource management block diagram in a cloud system is depicted in Fig. 1.2. It can be seen that clients are connected to a cloud server through a web portal. Users send their workloads to the cloud server for the execution. In turn, a modern cloud system tries to assign the workloads to one of the server machines based on different criteria including resource utilization, system performance, user priorities, operational cost, quality of service, etc. Typically, the complete process of workload placement over a time to improve different variables of a system is referred to as cloud resource management. As depicted in Fig. 1.2, the major tasks of a cloud resource management application are workload analysis and forecasting, resource provisioning, and scheduling the workloads on hardware. The workload analysis module is responsible for analyzing the upcoming workload and for forecasting the expected workload in the near future. This information is used by the resource provisioning module to allocate the physical resources. The resource scheduler places the workloads on the servers based on the input from the resource provisioning module and current resource usage information. Typically, resource management is achieved through prediction, scaling, provisioning, and load balancing, as shown in Fig. 1.3. However, this book concentrates on workload forecasting using different approaches of regression analysis and artificial neural networks, and load balancing.

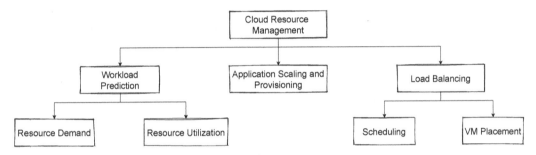

FIGURE 1.3 Cloud resource management approaches

1.2.1 Workload Forecasting

The workload prediction is a mechanism that estimates the future workload on the servers and can be classified as either a homeostatic or history-based method [89]. The first class of methods detect a trend in previous actual values and add or subtract it to the current value to forecast the next value. It could be a static or dynamic value that is to be added/subtracted. On the other hand, the second class of models analyzes the workload history and extracts a pattern to forecast the next value. A homeostatic method attempts to follow the mean of the previous values, while the history-based approach uses the behavior of previous workload information to forecast the next instance [70,114,116−118].

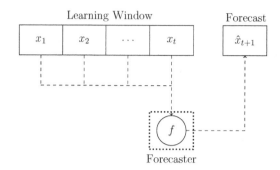

FIGURE 1.4 Schematic representation of workload forecasting

$$\hat{x}_{t+1} = f\left(x_t, x_{t-1}, \ldots, x_1\right) \tag{1.1}$$

Let f be a function of X which determines the value of $\hat{x}_{t+1}$ i.e. the estimated workload at time $t+1$. In order to forecast the upcoming workload, the f analyses the historical workloads of the length of the learning window (eq. 1.1). For instance, the function analyses the previous 10 instances from history if the length of the learning window is 10. Since the forecast function of a real data-trace is generally non-linear and complex, it becomes a challenging task to find the set of optimal parameters. And machine learning becomes the natural choice to optimize the model parameters to forecast the dynamic and non-linear workloads. A typical workload forecasting model is depicted in Fig. 1.4. The learning window defines the number of recent past workload instances to be analyzed for anticipation of next the value.

The prediction models have been explored and developed for various applications [13, 37, 46, 96, 97, 99]. The workload predictive resource management approaches are tailored with estimations of demand and utilization of resources. The workload of cloud services is dynamic and varies over time [30, 100]. Therefore a robust prediction model is required to produce reasonably accurate forecasts. On the other hand, the resource utilization prediction helps in accessing the free resources and also in accessing the impact of allocating the free resources to individual workloads [35, 76, 77, 79, 80, 115, 121, 123, 134].

1.2.2 Load Balancing

The task of a load balancing process is to distribute the workloads uniformly among servers. The load balancers are responsible for identifying the best suitable servers or computing resources that meet the application requirements. It ensures that the high volume of network traffic is not diverted to a single server. The schematic representation of load balancing in a distributed computing environment is shown in Fig. 1.5. The load balancer receives the traffic of users' requests through the Internet and distributes it among accessible and eligible servers. For instance, when seven users send their workloads, the load balancer balances the load distribution by assigning the load to three of the four servers.

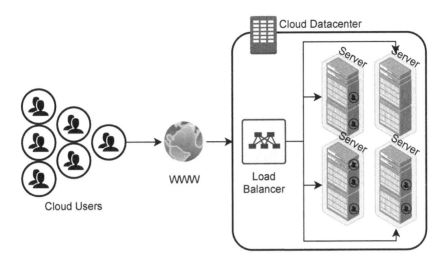

FIGURE 1.5 Load balancing

The effective load balancing is another approach that helps in achieving better usage of resources and their management. The efficiency of load balancing approaches has been an issue for cloud systems since its development [84,129]. The efficiency in load balancing can be achieved using different approaches such as optimal scheduling and placement of workloads or virtual machines (VMs). The optimal mapping of VMs is a complex and challenging task as it involves multiple objectives to optimize at the same time and belongs to NP-Complete class of problems [16,88]. Generally, the existing VM placement algorithms consider the different dimensions of resource utilization and power consumption in the data centers [4,138]. We will focus on the load balancing approaches, also dealing with the security while balancing the load on cloud servers as it is one of the most important issues in the cloud architectures, and various approaches have been discovered including [43,85].

1.3 MACHINE LEARNING

Machine learning allows a computer to master a specific task without being explicitly programmed. A computer can extract the underlying rules to perform the given task from a bunch of data points. In this book, we will discuss the cloud management models which are developed using the following techniques:

1.3.1 Artificial Neural Network

An artificial neural network (ANN) is composed of huge interconnected nodes called neurons, as shown in Fig. 1.6, that processes any information in a similar way as of human brain. Typically, a neural network is trained to solve a specific complex problem such as recognition, classification, forecasting, and others. Similar to the brain, a neural network adjusts the connections and their weights during the learning phase. The neural networks are capable of analyzing the complex and large amount of data and extracting the patterns from it.

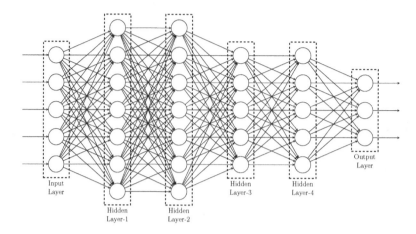

FIGURE 1.6 Artificial neural network

The key difference between a traditional computing approach and a neural network is that the traditional approach follows a set of rules that must be known to the computer in advance, while a neural network can learn from the data itself to draw insightful inferences using some specific rules. Let $\varkappa_1 = [x_1, x_2, \ldots, x_t]$ be an input vector, and the network, as shown in Fig. 1.6, is applied to estimate the value of x_{t+1}. Assuming that $\omega_{i,j}^k$ represents the weight of a synaptic connection between the ith node of the kth layer and the jth node of the next layer, and ζ_k denotes the activation function applied on k^{th} layer nodes. The output of the jth node of layer $k+1$ can be computed as $z_j = \sum_{i=1}^{t} \zeta_{k+1}(x_i \cdot \omega_{i,j}^k)$ that acts as the input to next layer nodes.

1.3.2 Metaheuristic Optimization Algorithms

Optimization has become an integral part of solving real-world problems which are multi-modal and highly non-linear in nature. These problems can be represented as a constrained optimization problem with one or more decision variables. Some of the real-world optimization problems are routing, engineering designs, resource assignment, and most of these problems are NP-hard [16, 86]. Consider that Fig. 1.7 shows an arbitrary function f_{arb} of decision variable x that needs to be minimized. The solution space has an infinite number of solutions along with multiple local optima such as $S_{l_1}, S_{l_2}, S_{l_3}$, and many others. If the solution space is multidimensional, complex, and large enough, it becomes a challenging task to find the global optimal (S_{g^*}) in a reasonable time. The metaheuristic optimization algorithm helps in solving such problems. The term *metaheuristic* was first used by Fred Glover [47] for an approach that has the capability of guiding and modifying the other heuristics to produce the solutions beyond their ability [48]. A metaheuristic algorithm does not guarantee to produce an optimal solution, but it generates an approximated solution in a reasonable amount of time.

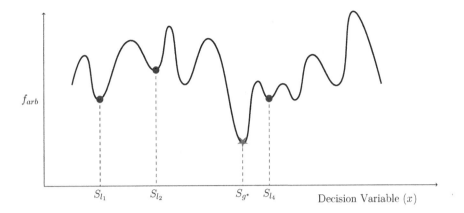

FIGURE 1.7 An arbitrary optimization function with multiple local optima

These algorithms can be classified into two major categories i.e. trajectory-based and population-based approaches. A trajectory-based algorithm such as Simulated Annealing works around a single solution to find an optimal solution for the problem under consideration. On the other hand, a population-based algorithm uses a set of solutions to search for an optimal solution. A detailed study on metaheuristic optimization can be seen in [14].

1.3.3 Time Series Analysis

A sequence of data points obtained at regular interval and indexed in time is referred to as time series data. Time series analysis started a long ago in 1927 [128] and has a range of applications including finance, signal processing, astronomy, forecasting, stock market, statistics, defense, politics, etc. In general, the task of a time series analysis model is to extract the meaningful data characteristics to predict the data trends. The time series models predict the future event after analyzing the historical events, and a number of models are introduced. Time series analysis is applicable to any kind of data including numeric, symbolic, continuous, and real-valued data.

1.4   WORKLOAD TRACES

The analysis reported in this book is carried out on various data traces. The workload data traces belong to two different categories, namely web server workloads and cloud server workloads.

HTTP-Web Server Logs: The HTTP traces of web servers of NASA, Calgary, and Saskatchewan servers are used [1]. In this book, these data traces are referred to as NASA Trace (D_1), Calgary Trace (D_2), Saskatchewan Trace (D_3), respectively. The D_1 is composed of two months of HTTP web requests obtained from the WWW server of NASA Kennedy Space Center in Florida. Similarly, the D_2 data-trace contains the HTTP request of one-year duration obtained from the WWW server located at the University of Calgary, Alberta, Canada. On the other hand, the D_3 is a data-trace that contains the HTTP server requests of seven months obtained from a WWW server of a university at Saskatchewan. Every data trace stores the records in ASCII files, and every line stores one record. Every record is composed of five records i.e. *host, timestamp, request, HTTP reply code*, and *bytes in the reply.*

Google Cluster Trace: It contains the data collected from the cluster cell of Google for 29 days of duration. The workload trace was released in 2011, and it contains the data from 10388 servers, 20 million tasks, and more than 0.67 million jobs [112]. A job is a set of one or more tasks, and tasks are further decomposed into one or more processes. In this book, the CPU and Memory resource demands are used and referred to as CPU Trace (D_4) and Memory Trace (D_5).

PlanetLab Trace: It is a collection of mean CPU utilization data which is collected from 11,746 virtual machines. These virtual machines are scattered at 500 different locations across the world. The data was collected for randomly selected 10 days during March and April of 2011, and data was sampled on five minutes intervals. The CPU utilization data of 22 randomly selected virtual machines is used in this book and referred to as PlanetLab Trace (D_6).

1.5 EXPERIMENTAL SETUP & EVALUATION METRICS

A machine equipped which contains two Intel® Xeon® E5-2630 v4 processors, and both processors run at the clock speed of 2.20 GHz. The machine is equipped with 128GB of main memory, and it operates on 64-bit windows servers 2012 R2. The predictive frameworks discussed in this book are evaluated using below mentioned metrics:

Mean Squared Error: The mean squared error (MSE) measures the forecast accuracy, and it is one of the popular metrics used in the literature. This method heavily penalizes the large error terms. Mathematically, it is denoted as given in eq. (1.2), where m represents the size of data in a given trace. The term MSE and MPE (mean squared prediction error) are interchangeably used in the book. Moreover, the square root of MSE (RMSE) may also be used as an error metric.

$$MSE = \frac{1}{m} \sum_{t=1}^{m} (x_t - \hat{x}_t)^2 \qquad (1.2)$$

Mean Absolute Error: A small number of very large magnitude errors may influence the accuracy measured using mean squared error. Whereas mean absolute error equally weights every error term, and it computes the mean of absolute differences between predicted and actual workloads as given in eq. (1.3). The forecasts are close to the actual workload values if the measured score is close to zero.

$$MAE = \frac{1}{m} \sum_{t=1}^{m} |x_t - \hat{x}_t| \qquad (1.3)$$

Relative Mean Absolute Error: A scale-free error metric is required to compare the forecast models on different data sets, and relative mean absolution error (RelMAE) is one such metric. The score can be calculated using eq. (1.4), which represents the mean absolute error of the algorithm (MAE_A) normalized by the mean absolute error of a base or state of the art model (MAE_{BM})

$$RelMAE = \frac{MAE_A}{MAE_{BM}} \tag{1.4}$$

Mean Absolute Scaled Error: Rob J. Hyndman and Anne B. Koehler introduced a new metric as a substitution of percentage error metrics [61]. The prediction errors are scaled on the basis of the training mean absolute error of a naïve forecast method. It computes the measured score using eq. (1.5), where m_s denotes the seasonal term. This metric is a good choice of accuracy measurement when the prediction model is compared across a number of different scales.

$$MASE(x, \hat{x}) = \frac{1}{m} \sum_{t=1}^{m} \left(\frac{|x_t - \hat{x}_t|}{\frac{1}{m-m_s} \sum_{t=m_s+1}^{m} |x_t - x_{t-1}|} \right) \tag{1.5}$$

Correlation Coefficient: The correlation coefficient (CoC) statistically evaluates the statistical relationship of two variables by measuring the degree of movements. The CoC score can be computed as given in eq. (1.6), where $\bar{x}$ and $\bar{\hat{x}}$ are the mean values of actual and predicted workloads, respectively.

$$CoC_{x\hat{x}} = \frac{\sum (x_t - \bar{x})(\hat{x}_t - \bar{\hat{x}})}{\sqrt{\sum (x_t - \bar{x})^2 \sum (\hat{x}_t - \bar{\hat{x}})^2}} \tag{1.6}$$

Sum of Elasticity Index: Messias et al. proposed to use the sum of elasticity index (SEI) as a measure of forecast accuracy [92]. This metric supports a forecast model having the best performance most of the time. As opposed to MAE and RMSE, it is very less sensitive to the outliers. The SEI score is computed as given in eq. (1.7) and it always lies between zero and one, where zero and one define the worst and best accuracy of the model.

$$SEI = \sum_{t=1}^{m} \frac{min(x_t, \hat{x}_t)}{max(x_t, \hat{x}_t)} \tag{1.7}$$

1.6 STATISTICAL TESTS

The statistical techniques are used to analyze the forecasting behavior of different approaches. The non-parametric tests are used due to the fact that they are not highly restrictive and can be used over small sample sizes [44]. The significance tests help in finding the presence of significant differences in two or more forecasting models.

1.6.1   Wilcoxon Signed-Rank Test

The Wilcoxon signed-rank test is one of the non-parametric tests that compare two samples to find out whether they represent the same population or not [130]. It assumes a null hypothesis (H_0^{WC}) that the mean of both samples is the same. If a significant difference is detected, H_0^{WC} gets rejected.

Considering that two algorithms are being evaluated on k problems and ϵ_i denotes the performance score difference on the ith problem. The method ranks the absolute differences and computes the ranks accordingly. The ties can be addressed using one of the available approaches. In this book, the number of ties is equally divided to compute the rank of both algorithms. The total rank of the first algorithm where it outperforms the second is computed using eq. (1.8) whereas eq. (1.9) computes the sum of ranks for the problems where the second algorithm gives better results than the first algorithm.

$$\mathrm{R}_{\mathrm{WC}}^{+} = \sum_{\epsilon_i > 0} rank(\epsilon_i) + \frac{1}{2} \sum_{\epsilon_i = 0} rank(\epsilon_i) \qquad (1.8)$$

$$\mathrm{R}_{\mathrm{WC}}^{-} = \sum_{\epsilon_i < 0} rank(\epsilon_i) + \frac{1}{2} \sum_{\epsilon_i = 0} rank(\epsilon_i) \qquad (1.9)$$

1.6.2   Friedman Test

It is a non-parametric test developed by M. Friedman [40, 41] that provides an alternative to one-way ANOVA with repeated measures. It conducts multiple tests that target to detect the presence of differences between the performance behavior of two or more models [34].

Let H_0^{FR} be the null hypothesis of Friedman test that states the equality in the mean of every prediction model's result. The alternate hypothesis (H_1^{FR}) of the test is the negation of H_0^{FR}. First, it converts the original results of each algorithm into ranks. Let $\mathrm{R}_{\mathrm{FR}}^{\epsilon_i^j}$ be the Friedman rank of jth algorithm on the ith problem, the final rank of jth algorithm can be observed by calculating the average of $\mathrm{R}_{\mathrm{FR}}^{\epsilon_i^j}$ as shown in eq. (1.10), where $j = \{1, 2, \ldots, k\}$ and $i = \{1, 2, \ldots, |D|\}$ denote the algorithms and datasets respectively, $|D|$ is the number of datasets. The minimum value of ranks represents the best algorithm.

$$\mathrm{R}_{\mathrm{FR}}^{\epsilon^j} = \sum_{i=1}^{|D|} \mathrm{R}_{\mathrm{FR}}^{\epsilon_i^j} \qquad (1.10)$$

1.6.3   Finner Test

The Friedman test conducts the multiple comparison test and detects the significant difference over the whole population test. However, it is unable to conduct comparisons to detect the difference between some of the algorithms. Post-hoc analysis tests deliver the purpose and allow to detect the presence of difference in the performance of two algorithms on the basis of a control method [38]. The test adjusts the value of significance level ($\aleph$) in a step-down manner [34]. Considering that the generated p-values are sorted in an increasing fashion in such a way that $p_i \leq p_{i+1}; \forall i = \{1, 2, \ldots, k-2\}$. Let H_i^{FN} be the corresponding hypothesis for tests. The Finner test rejects the hypothesis from H_1^{FN} to H_{i-1}^{FN} provided i is the smallest integer number that satisfy $p_i > 1 - (1-\aleph)^{\frac{k-1}{i}}$ property [34].

CHAPTER 2

Time Series Models

TIME SERIES analytical models are being used in forecasting since a long ago in 1927 [128]. A time series-based model forecasts the trends after analyzing the various characteristics of data indexed in time. Since their first usage, they have been widely used in scientific research and industry-oriented applications. This chapter concentrates on univariate time series-based workload forecasting. A univariate time series can be defined as a collection of measurements of the same variable over time (typically, at regular time intervals). The essential characteristic of any time series data is that the order of observation matters and change in order may alter the significance of the data.

The time series analysis is typically associated with the process of finding a model to fit the time series data. The observed model can be used to extract the pattern, forecast future events, and explain the effects of past events on the future. Some of the essential characteristics of a time series are:

- The *trend* depicts the direction of the data i.e. to increase or decrease. The direction is always need not be in the same direction for a long period of time. According to the Organisation for Economic Co-operation and Development (OECD), "The trend is the component of a time series that represents variations of low frequency in a time series, the high and medium frequency fluctuations having been filtered out."

- The *seasonality* also depicts similar characteristics as of trend. The difference between the two terms is that the seasonality shows repetitive patterns.

- The *noise* referred to the component depicting neither trend nor seasonality in the data.

- *Outliers* are the far-away data points from the data.

- The other common characteristics in a time series data are *long-run cycle, constant variance over time, and spikes.*

This chapter discusses the five basic time series analysis models and uses them to forecast the different types of workloads on cloud servers. A detailed analysis is conducted to validate their performance on real-world data traces.

DOI: 10.1201/9781003110101-2

2.1 AUTOREGRESSION

An autoregressive model is used to present a phenomenon where the future values of any variable are the function of its historical values. It is used to depict a random process in real-world applications of statistics, signal processing, data analysis, etc. Formally, an autoregressive model can be defined as a process that considers the historical data with a white noise term to generate the future outcome of a variable. This model gets the name from its functioning as it regresses the same variable. Considering that the cloud workload x is indexed in equally spaced time interval $1, 2, \ldots, t$ as $x_1, x_2, \ldots, x_t$ then the autoregressive model of order p can be defined as eq. (2.1) [91].

$$\hat{x}_t = \phi_1 \times x_{t-1} + \phi_2 \times x_{t-2} + \ldots + \phi_p \times x_{t-p} + \aleph_t \tag{2.1}$$

The $\hat{x}_t$ and x_t are predicted and actual workloads respectively at time t; and model parameters are represented as ϕ_i $(i = 1, 2, \ldots, p)$ which should hold any value from -1 to +1. The term $\aleph_t$ is used to denote the white random noise. The working of an autoregressive model is graphically shown in Fig. 2.1.

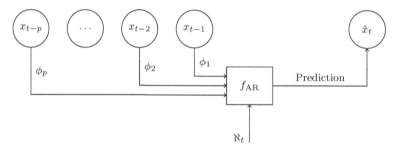

FIGURE 2.1 Autoregression process

2.2 MOVING AVERAGE

The moving average (MA) model uses the errors in the previous forecasts rather than the previous actual values. According to Box et al., the time series can be modeled using Ξ if successive actual values of the series are highly correlated [15]. These error terms can be considered as the zero-mean white noise series generated using a fixed distribution. A typical moving average model of order q can be written as eq. 2.2, where θ_j represents the weight for j^{th} model term. These weights are required neither to be positive nor to be total unity [15]. The graphical illustration of a moving average model with q model terms is shown in Fig. 2.2, where f_{MA} denotes the function that learns the moving average model.

$$\hat{x}_t = \theta_1 \times \xi_{t-1} + \theta_2 \times \xi_{t-2} + \ldots + \theta_q \times \xi_{t-q} + \aleph_t \tag{2.2}$$

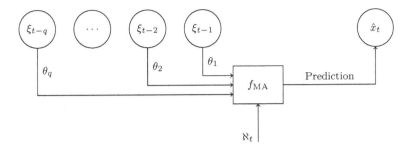

FIGURE 2.2 Moving average process

2.3 AUTOREGRESSIVE MOVING AVERAGE

The autoregressive moving average (ARMA) model is a combination of two different forecasting models. It suggests using the actual historical values along with the errors associated with the previous forecasts. The ARMA process can be represented mathematically as shown in eq. 2.3. The ARMA model parsimoniously represents time-series data using two different polynomial fits associated with the auto regression and moving average models respectively. For a complicated time series data, the ARMA process is preferred over autoregression and moving average models. The graphical representation of the ARMA model is shown in Fig. 2.3 having p and q model terms for autoregression and moving average models.

$$\hat{x}_t = \phi_1 \times x_{t-1} + \phi_2 \times x_{t-2} + \ldots + \phi_p \times x_{t-p} + \aleph_t + \theta_1 \times \xi_{t-1} + \theta_2 \times \xi_{t-2} + \ldots + \theta_q \times \xi_{t-q} \quad (2.3)$$

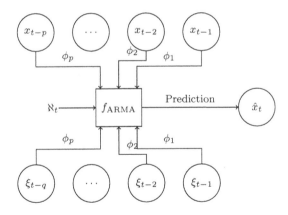

FIGURE 2.3 Autoregressive moving average process

2.4 AUTOREGRESSIVE INTEGRATED MOVING AVERAGE

The more general form of an ARMA process is the autoregressive integrated moving average, commonly referred to as ARIMA. This model is composed of three components viz. Autoregression (AR), Integration (I), and Moving Average (MA). The ARIMA process is a good choice to model non-stationary time series. The ARIMA process integrates the non-stationary data to transform into stationary data by applying the difference operator. The difference operator is depicted as δ^d, where d is the number of different terms. For instance, the first-order difference operation can be defined as $\delta^1 = x_t - x_{t-1}$. The graphical representation of an ARIMA(p, d, q) is depicted in Fig. 2.4, where p, d, and q are the model terms or weights associated with autoregression, difference, and moving average respectively. The workload values at time t obtained after applying the difference are denoted as $\tilde{x}_t$. The first order ARIMA is one of the simplest models that can be represented as shown in eq. 2.4, where the term B denotes the backward shift operator ($B \times x_t = x_{t-1}$) [91].

$$\underbrace{(1 - \theta_1 B)}_{\text{AR}(1)} \underbrace{(1 - B)x_t}_{\text{First Difference}} = \underbrace{(1 - \theta_1 B)\xi_t}_{\text{MA}(1)} \tag{2.4}$$

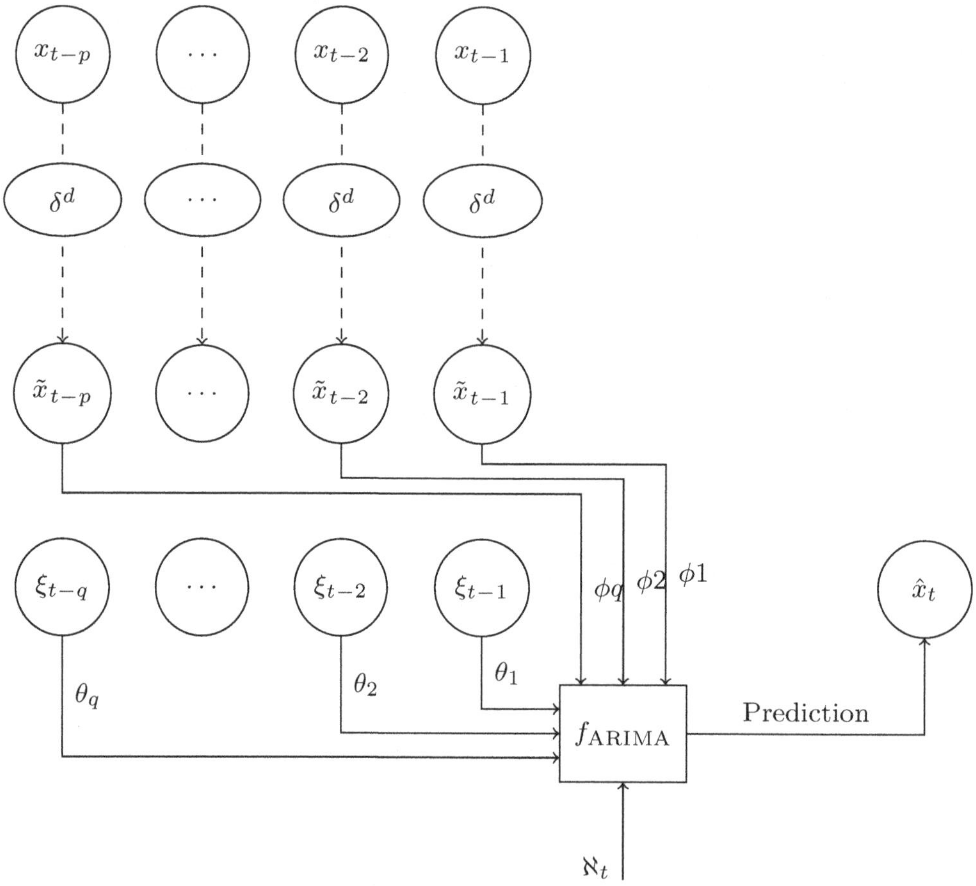

FIGURE 2.4 Autoregressive integrated moving average process

2.5 EXPONENTIAL SMOOTHING

As opposed to the regression-based time series analysis models, the exponential smoothing (ES) model uses the historical actual values and their corresponding forecasts. This model suggests that the recent predictions are highly important to consider for better modeling of time series data. The term exponential comes from the principle of exponentially reducing the weights associated with the previous forecasts. The mathematical representation of the model is shown in eq. 2.5, where the term α is associated with the smoothing constant. The exponential smoothing is graphically illustrated in Fig. 2.5. One of the obvious characteristics of the model is that it does not need to store a large amount of historical values as oppose to the regression-based time series models.

$$\hat{x}_t = \alpha x_{t-1} + (1 - \alpha)\hat{x}_{t-1} \tag{2.5}$$

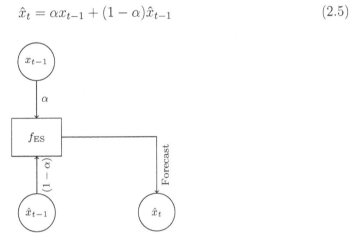

FIGURE 2.5 Exponential smoothing process

2.6 EXPERIMENTAL ANALYSIS

The performance of these models on forecasting the cloud server workloads is assessed with a number of experiments. The workload is forecasted with different values of the prediction window size. The term PWS defines the time interval in two consecutive forecasts, for instance, if a model estimates the upcoming workload for every 60 minutes then the length of the prediction window size is 60 minutes. This study is conducted with the length of the prediction window of size 5, 10, 20, 30, and 60 minutes duration. Also, the model parameters are estimated using 60% of the data and the remaining 40% of the data is used to observe the accuracy of the time series models with estimated parameter values.

2.6.1 Forecast Evaluation

The forecast accuracy is measured using mean absolute error (MAE) and mean absolute scaled error (MASE). The data sets are categorized into two categories viz. web server workloads and cloud server workloads. The forecast accuracy on web server workloads is shown in Figs. 2.6a and 2.7a using MAE and MASE respectively. The model forecasts Calgary Trace with the least MAE for the prediction window of length 5, 10, 20, and 40 minutes. For the prediction window of 60 minutes duration, the least mean absolute error is obtained on the forecasts of Saskatchewan Trace. Similarly, the MASE-based forecast results are depicted in Figs. 2.6a and 2.7b for web and cloud server workloads respectively. It is evident that the D_2 forecasts are most accurate for the length of the prediction window of 5, 10, 30, and 60 minutes as per the MASE. For 20 minutes of the duration, the best forecast accuracy measured using MASE is generated for D_1. Based on the results, the first-order autoregressive process models the web server workloads with better accuracy than cloud server workloads. For cloud server-based workloads, the memory trace (D_5) obtained better results over CPU trace. In general, the first-order autoregressive process learns the pattern from Calgary trace in a better way.

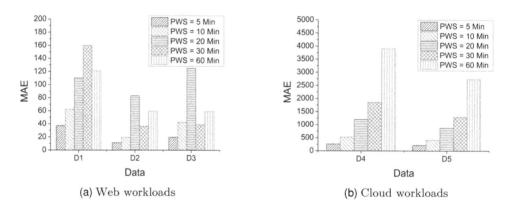

(a) Web workloads   (b) Cloud workloads

FIGURE 2.6 Autoregression forecast results on MAE

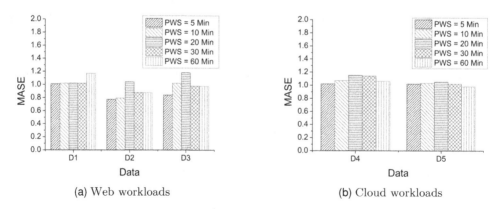

(a) Web workloads   (b) Cloud workloads

FIGURE 2.7 Autoregression forecast results on MASE

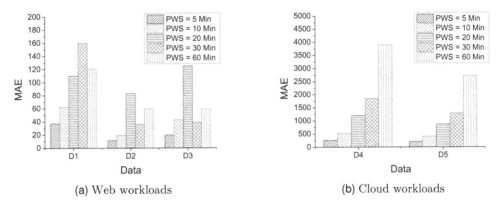

(a) Web workloads   (b) Cloud workloads

FIGURE 2.8 Moving average forecast results on MAE

For moving average model also, the experiments are conducted with the same settings as of autoregression. Again, the first-order moving average model is used for the study to ensure similarity and simplicity. The MAE-based forecast results are shown in Figs. 2.8a and 2.8b. It is evident from the results that the model learns the pattern of Calgary trace with more accuracy over other data sets. In overall comparisons, it is observed that the MA(1) model is more suitable to the web workloads due to the fact that the model attained large errors while forecasting the workload on cloud servers. Among the cloud workloads, the model generates better forecasts for memory workload over the other trace. The forecast results obtained on MASE are shown in Figs. 2.9a and 2.9b for web and cloud server workloads respectively. A similar trend is observed in the assessment of MASE-based experiments.

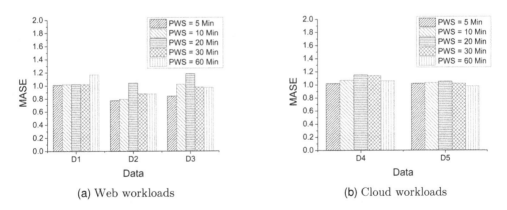

(a) Web workloads   (b) Cloud workloads

FIGURE 2.9 Moving average forecast results on MASE

For ARMA process also, the experiments are done with first-order ARMA i.e. ARMA(1,1). The forecast results based on MAE for web and cloud server workloads are shown in Figs. 2.10a and 2.10b respectively. While the forecast results obtained on MASE are depicted in Figs. 2.11a and 2.11b respectively. The model has successfully

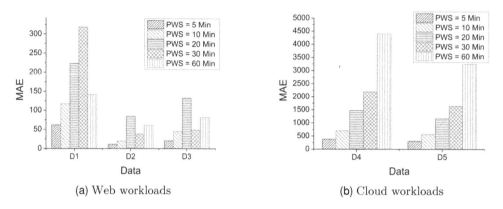

(a) Web workloads   (b) Cloud workloads

FIGURE 2.10 Autoregressive moving average forecast results on MAE

obtained better forecasts for Calgary trace on the prediction window size of 5, 10, and 20 minutes. For 30 and 60 minutes prediction intervals, the model is able to generate better forecasts for Saskatchewan trace. The similar trends are observed for MASE-based assessment. Moreover, the prediction error based on MASE is low for Calgary trace in 20 minutes. The ARMA model shows a similar trend on web and cloud server workloads.

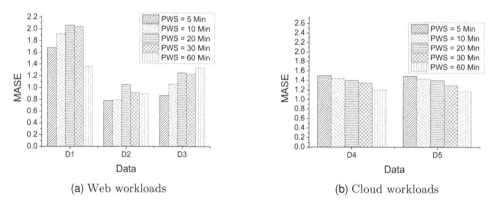

(a) Web workloads   (b) Cloud workloads

FIGURE 2.11 Autoregressive moving average forecast results on MASE

The performance of the ARIMA process is also assessed on the same experiments and the results are depicted in Figs. 2.12 and 2.13 for MAE and MASE respectively. The Calgary trace forecasts are obtained with the least prediction error. On the other hand, the MASE-based results advocate that the Saskatchewan trace and NASA trace are forecasted with the least error while the prediction window size is 5 and 20 minutes duration.

Unlike the regression-based models' forecast results, the exponential smoothing process does not model any single workload for different prediction window sizes.

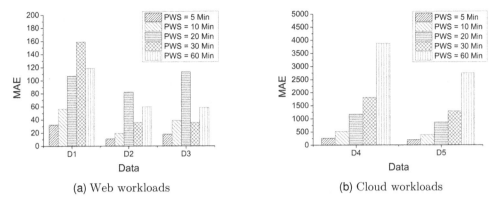

FIGURE 2.12 Autoregressive integrated moving average forecast results on MAE

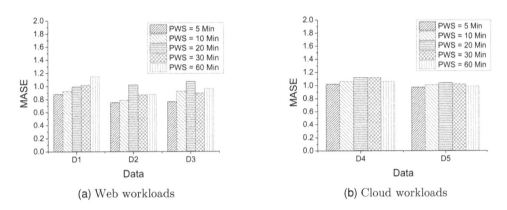

FIGURE 2.13 Autoregressive integrated moving average forecast results on MASE

For different duration of prediction window sizes, the least forecast error is obtained for different workload traces. For instance, during the prediction window size of 5, 10, and 60 minutes, and in other cases, the D_1 results are better, according to the MAE-based results shown in Fig. 2.14. The MASE forecast errors are shown in Fig. 2.15 and it shows that the D_2 trace forecast errors are low during the 10 and 30 minutes prediction window. In the case of a 20-minute prediction interval, the model generates the best forecasts for the Saskatchewan trace. Similarly, the CPU trace is best modeled during a 60-minute prediction window. The D_2 and D_3 attained the least and equal forecast accuracy during 5-minute intervals.

2.6.2 Statistical Analysis

The experimental observations find that each data set is modeled with different related forecast accuracy by all-time series models. Thus, it becomes a challenging task to conclude the rankings of the models or to advocate the use of a model for given data traces. Therefore, a deep investigation is carried out using statistical analysis which helps in finding the significant difference in the performance of different models. For this purpose, the Friedman test [40] is applied because this test is believed to be one of the most powerful tests if the sample size is five or more [28]. A null hypothesis (H_0^{FR}) is followed which states that there is no significant difference in the performance and each model behaves similarly. On the other hand, the alternate hypothesis (H_1^{FR}) is the negation of the null hypothesis i.e. the performance of one of the models is significantly different than others.

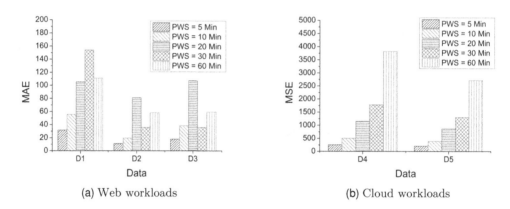

(a) Web workloads   (b) Cloud workloads

FIGURE 2.14 Exponential smoothing forecast results on MAE

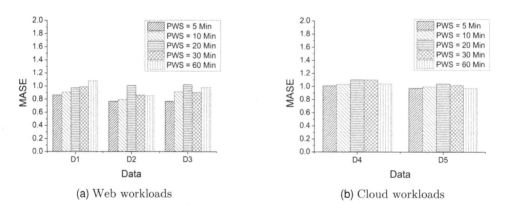

(a) Web workloads   (b) Cloud workloads

FIGURE 2.15 Exponential smoothing forecast results on MASE

The STAC web platform [113] is used to conduct a detailed statistical analysis. The corresponding F-statistics and p-values are listed in Table 2.1, where the rejection of null hypothesis (H_0^{FR}) is represented by H_0^{FR}.R. The algorithms are ranked by the test as per the difference in their performance and the rank results based on both MAE and MASE are shown in Table 2.2 for $\aleph = 0.05$. The lowest value indicates the highest rank and vice-versa.

TABLE 2.1 Friedman test statistics for time series forecasting models

Accuracy Metric	F-Statistics	p-value	H_0^{FR} Result
MAE	45.42	0.0	H_0^{FR}.R
MASE	34.70	0.0	H_0^{FR}.R

TABLE 2.2 Friedman test ranks for time series forecasting models

Model	MAE	MASE
AR (M_1)	3.60	3.54
MA (M_2)	4.94	4.86
ARMA (M_3)	2.44	2.42
ARIMA (M_4)	2.28	2.38
ES (M_5)	1.74	1.80

TABLE 2.3 Finner test post-hoc analysis of time series forecasting models

Comparison	MAE			MASE		
	Statistic	p-value	H_0^{FN} Result	Statistic	p-value	H_0^{FN} Result
M_1 vs M_2	2.996	0.00546	H_0^{FN}.R	2.951	0.00631	H_0^{FN}.R
M_1 vs M_3	2.593	0.01353	H_0^{FN}.R	2.504	0.01748	H_0^{FN}.R
M_1 vs M_4	2.951	0.00546	H_0^{FN}.R	2.593	0.01577	H_0^{FN}.R
M_1 vs M_5	4.159	0.00008	H_0^{FN}.R	3.890	0.00025	H_0^{FN}.R
M_2 vs M_3	5.590	0.00000	H_0^{FN}.R	5.545	0.00000	H_0^{FN}.R
M_2 vs M_4	5.947	0.00000	H_0^{FN}.R	5.545	0.00000	H_0^{FN}.R
M_2 vs M_5	7.155	0.00000	H_0^{FN}.R	6.842	0.00000	H_0^{FN}.R
M_3 vs M_4	0.357	0.72051	H_0^{FN}.A	0.089	0.92873	H_0^{FN}.A
M_3 vs M_5	1.565	0.14468	H_0^{FN}.A	1.386	0.20257	H_0^{FN}.A
M_4 vs M_5	1.207	0.24907	H_0^{FN}.A	1.296	0.21380	H_0^{FN}.A

It is evident that the test does not accept the null hypothesis which means there are at least one of the algorithms whose results are significantly different. However, the Friedman test is incapable of further investigation on which model is better or so. Fortunately, the post-hoc analysis does the job and gives a detailed analysis. This study uses the Finner post-hoc test [38] which conducts multiple comparisons. One of the methods is selected as the control method and compares its performance with every other model. This test also works with a null hypothesis represented as H_0^{FN} which assumes the equality between the mean of the results of the selected control method and every other group member participating in the test. The results of the pairwise comparisons conducted by the Finner test are shown in Table 2.3, where rejection and acceptance of the null hypothesis are represented using H_0^{FN}.R and H_0^{FN}.A respectively, and $M_1, M_2, M_3, M_4,$ and M_5 denote AR, MA, ARMA, ARIMA, and ES processes respectively. The test does not accept the null hypothesis in most of the paired comparisons that indicate the presence of a significant difference in the performance of compared models. The statistical observations recommend a similar forecast accuracy of ARMA, ARIMA, and ES as the post-hoc analysis does not find any significant difference in their results. The Friedman test results also support the above observation as the difference in the ranking of these models is very low. However, the ES process has produced the best forecasts measured using both error metrics as the model has achieved the best ranking which supports the fact of using both historical actual and forecast values to predict the future event. The moving average is the worst performing model as it receives the lowest rank in the test.

CHAPTER 3

Error Preventive Time Series Models

T IME SERIES analysis is a popular choice to forecast future events if historical data indexed in time is available. These models are good in learning and extracting the patterns from the time-series data. However, they need to tune the parameters periodically to model the behavioral changes in data [6]. Dynamic systems such as cloud servers usually see frequent changes in the data pattern. Thus, time series analysis models may not forecast the cloud workloads with reasonable accuracy. An error prevention method is introduced to address the issue of periodic parameter tuning [78]. This method is generic in nature and can be integrated with any time series analysis model. It also improves the pattern capturing ability of a forecasting model.

3.1   ERROR PREVENTION SCHEME

The error prevention scheme aims to improve the pattern learning ability of a forecasting model. The recent forecasts are analyzed to compute the residual trend which is further used as an input for the next forecast. Let $X = \{x_1, x_2, \ldots, x_t\}$ be the set of workload values indexed in time and $\hat{X} = \{\hat{x}_1, \hat{x}_2, \ldots, \hat{x}_t\}$ is a set of corresponding forecasts obtained by a forecasting function given in eq. (3.1). The forecast error (ξ_t) is the difference between x_t and $\hat{x}_t$. Further, these errors are used to compute the *error preventive score* (EPS) at time t (EPS$_t$) using eq. (3.2), where the value of k is the length of the feedback window which is defined by the user. The operational summary of the model is shown in Fig. 3.1.

$$\hat{x}_t = f(x_{t-1}, x_{t-2}, \ldots, x_{t-n}) \tag{3.1}$$

$$EPS_t = \frac{1}{k}(\xi_{t-1} + \xi_{t-2} + \ldots + \xi_{t-k})$$

$$= \frac{1}{k}\sum_{i=1}^{k}\xi_{t-i} \tag{3.2}$$

DOI: 10.1201/9781003110101-3

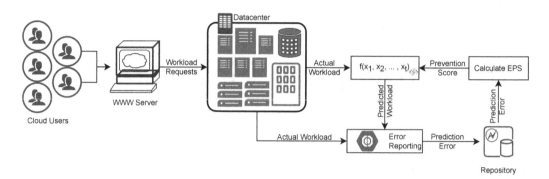

FIGURE 3.1 Error preventive workload forecasting model

Let $X = \{0.8147, 0.9058, 0.1270, 0.9134, 0.6324, 0.0975, 0.2785, 0.5469, 0.9575,$ $0.9649, 0.1576, 0.9706, 0.9572, 0.4854, 0.8003, 0.1419, 0.4218, 0.9157, 0.7922, 0.9595\}$ and $\hat{X} = \{*, 0.9755, 0.9974, 0.8105, 0.9992, 0.9318, 0.8034, 0.8468, 0.9113, 1.0098,$ $1.0116, 0.8178, 1.0129, 1.0097, 0.8965, 0.9721, 0.8141, 0.8812, 0.9998, 0.9701 \}$ are randomly selected data points and their corresponding forecasts obtained by a random function shown in eq. 3.3. The forecasted value at time $t = 1$ is not available due to non availability of x_0 and denoted as $*$.

$$\hat{x}_t = 0.78 + 0.24 \times x_{t-1} \tag{3.3}$$

The corresponding prediction errors are ($\Xi = \{*$, -0.0697, -0.8704, 0.1029, -0.3669, -0.8342, -0.5249, -0.3000, 0.0463, -0.0449, -0.8540, 0.1528, -0.0558, -0.5243, -0.0962, -0.8302, -0.3923, 0.0345, -0.2076, -0.0106\}). The error prevention score is computed using $k = 4$ as mentioned in eq. (3.4). The modified forecasting model that we referred to as error preventive model is obtained by including the error prevention score as shown in eq. 3.5.

$$EPS = \frac{1}{4} \sum_{i=1}^{4} \xi_i \tag{3.4}$$

$$\hat{x}_t = 0.9 + 0.1 \times x_{t-1} + \frac{1}{3} \sum_{i=t-1}^{t-3} \xi_i \tag{3.5}$$

After employing the error prevention score in the forecasting model, the new forecasts become $\hat{X}^{EP} = \{*, 0.9755, 0.9974, 0.8105, 0.9992, 0.6308, 0.3865, 0.6206,$ $0.6408, 0.9102, 1.0590, 0.6669, 0.9564, 0.8742, 0.6501, 0.9886, 0.5430, 0.5796, 0.8794,$ $0.7904\}$. The corresponding forecast errors are $\Xi^{EP} = \{*$, -0.0697, -0.8704, 0.1029, -0.3668, -0.5333, -0.1080, -0.0736, 0.3167, 0.0547, -0.9014, 0.3037, 0.0008, -0.3888, 0.1502, -0.8467, -0.1212, 0.3361, -0.0872, 0.1691\}. Figure 3.2 plots the actual workloads along with the forecasts generated by both original and modified prediction functions given in eqs. (3.1) and (3.5). It is evident that error preventive (EP) forecasts are closer to the actual workloads as opposed to the non-error preventive (NEP) forecasts. Thus, the error preventive score is useful to reduce the error in forecasts.

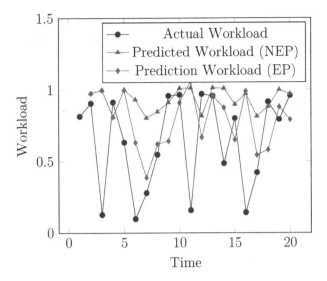

FIGURE 3.2 An example of error preventive and non-error preventive forecast

The order of execution time and required memory of a model is defined by its complexity. Any modification in the model will also change its complexity. Similarly, the inclusion of error prevention scores in the forecasting model also affects the complexity of the original model. Assuming that the k forecast errors which are used in computing the error prevention score are stored in an array of length k. Since computing the average of k values of an array consumes $\mathcal{O}(k)$, the inclusion of an error preventive score does not affect the complexity of a forecasting model by a large. In fact, the complexity of a predictive model will increase by a factor of $\mathcal{O}(k)$ only. Since the value of k is a very small number and can be considered as a constant factor. Therefore, the complexity of the updated model after including the error prevention score will remain as its original complexity provided that the forecasting model's complexity is more than $\mathcal{O}(k)$. Similarly, an array of length k needs $\mathcal{O}(k)$ memory space for storage. Thus the space complexity of a predictive model also gets increased by a factor of $\mathcal{O}(k)$ only.

3.2 PREDICTIONS IN ERROR RANGE

The prediction errors can be categorized and arranged into different categories or ranges as per the error margin. The 'predictions in error range' metric compute the ratio between the magnitude of the forecast and actual workload which is further multiplied with 100 to compute the percentage error as shown in eq. (3.6), where ξ_t^p denotes the percentage prediction error in $\hat{x}_t$. The magnitude of the error may lie in the range of 0 to ∞ which implies that $0 \leq \xi_t^p < \infty$. Five different categories are suggested to create the clusters of forecasts with different values as shown in Fig. 3.3. The number of categories is user input and can vary accordingly. The forecasts having ξ_t^p up to 25% are kept in the first category. The second, third, fourth, and fifth categories keep the records of forecasts having ξ^p values as shown in eq. (3.7). With this assignment each forecast gets associated with one of the categories. The number of data points belonging to each category is counted and stored in a variable PER_{r_i} as shown in eq. (3.8), where $\xi^p(r_i)$ is an array containing the forecasts belonging to r_i. This process is followed by the computation of a share of predictions for each category which is denoted as $\text{PER}_{r_i}^s$ (see eq. (3.9)). A model can be considered better if it increases $\text{PER}_{r_i}^s$ and reduces $\text{PER}_{r_j}^s$, where $i < j$.

$$\xi_t^p = \frac{|x_t - \hat{x}_t|}{x_t} \times 100 \qquad (3.6)$$

FIGURE 3.3 Prediction error range and segments

$$
\begin{aligned}
\xi_t^p \in r_1 && 0 \leq \xi_t^p \leq 25 \\
\xi_t^p \in r_2 && 25 < \xi_t^p \leq 50 \\
\xi_t^p \in r_3 && 50 < \xi_t^p \leq 75 \\
\xi_t^p \in r_4 && 75 < \xi_t^p \leq 100 \\
\xi_t^p \in r_5 && 100 < \xi_t^p < \infty
\end{aligned} \qquad (3.7)
$$

$$\text{PER}_{r_i} = count(\xi^p(r_i)) \qquad (3.8)$$

$$\text{PER}_{r_i}^s = \frac{\text{PER}_{r_i}}{\sum_{i=1}^{5} \text{PER}_{r_i}} \times 100 \qquad (3.9)$$

$$\text{where} \sum_{i=1}^{5} \text{PER}_{r_i}^s = 100$$

3.3 MAGNITUDE OF PREDICTIONS

In this competitive era, a cloud service provider would always prioritize the user experience to maintain its customer base. In this case, a cloud resource management application can tolerate the workload prediction more than the corresponding actual workload. If a lot of lower predictions (less than corresponding actual workloads) are produced, it would be more difficult to improve the user experience because as per the predictions less resources may be reserved to handle the future workload. Whereas the higher forecasts (more than the corresponding actual workload) will suggest reserving more active resources than required which can be helpful in improving the user experience by allowing a service provider to keep itself ready to fulfill all incoming requests without any wait on the cost of increased operational cost (in comparison to the optimal operational cost). Thus, a predictive model which generates more higher forecasts is preferred for the applications such as cloud resource management.

The 'magnitude of forecasts' (MoP) measures the accuracy of a predictive model on the basis of the magnitude of the forecast error. If a forecast $(\hat{x}_t)$ is higher than the corresponding x_t, the model assigns the data sample to a cluster of positive or higher forecasts (MoP$^+$) as shown in eq. (3.10). Whereas a forecast with negative magnitude error is assigned to a cluster of lower predictions (MoP$^-$). Further, the predictive models are assessed by comparing the number of forecasts in both clusters. Here, a predictive model with more forecasts in MoP$^+$ cluster is considered better.

$$MoP_t = \begin{cases} MoP^+ & \text{if } \hat{x}_t > x_t \\ MoP^- & \text{if } \hat{x}_t < x_t \end{cases} \tag{3.10}$$

3.4 ERROR PREVENTIVE TIME SERIES MODELS

In this section, we will discuss the time series models equipped with error prevention schemes, and their performance is compared with non-error preventive time series models.

3.4.1 Error Preventive Autoregressive Moving Average

An ARMA process simply combines two basic time series processes viz. autoregression and moving average. Thus, it uses two different polynomials for autoregression and moving average respectively to model the time series data as shown in eq. (2.3). A workload forecasting model (WP$_{\text{ARMA}}$) which uses an ARMA(1,1) can be represented as shown in eq. (3.11) and it is being referred to as $\mathcal{WP}_{\text{ARMA}}$ in this chapter.

$$\hat{x}_t = \phi_1 \times x_{t-1} + \theta_1 \times \xi_{t-1} + \aleph_t \tag{3.11}$$

An ARMA process enabled with the error prevention score is referred to as error preventive ARMA ($\mathcal{WP}_{\text{ARMA}}^{\mathcal{EP}}$) as illustrated in eq. (3.12). A set of experiments are conducted on five data sets (D$_1$, D$_2$, D$_3$, D$_4$, and D$_5$) with different values of the prediction window. The performance of the models is assessed using five different metrics viz. R-Score, SEI (Sum of Elasticity Index), MPE (Mean Prediction Error), Predictions in Error Range (PER), and Magnitude of Predictions (MoP).

$$\hat{x}_t = \phi_1 \times x_{t-1} + \theta_1 \times \xi_{t-1} + \aleph_t + EPS_t \tag{3.12}$$

The choice of error feedback window length is very important. Therefore, a critical analysis is conducted to choose a suitable length of the feedback window. The performance of both models (error preventive and non-error preventive) is assessed on different lengths of error feedback window i.e. $k = \{0, 5, 10, \ldots, 1000\}$. The forecast results for all data traces for prediction windows of sizes 10, 20, 30, and 60 minutes are depicted in Figs. 3.4, 3.5, 3.6, and 3.7 respectively. A lower value of error feedback window length is a preferable value as indicated by the observed results. Thus, the performance of both models is further compared on five terms of the feedback window length.

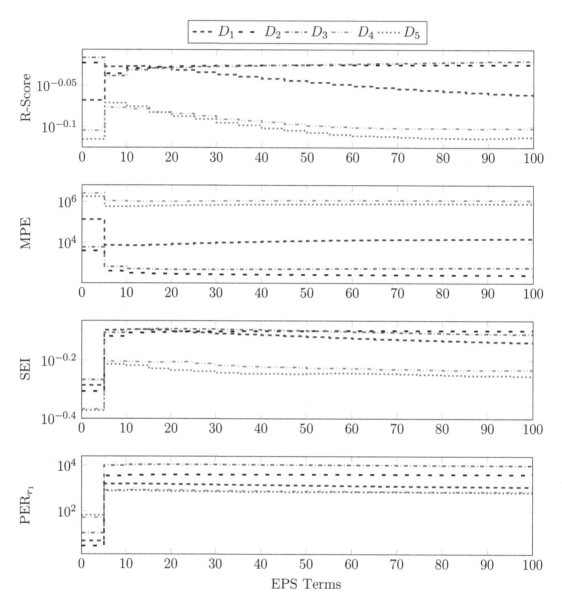

FIGURE 3.4 Error preventive ARMA forecast analysis for 10-minute prediction interval

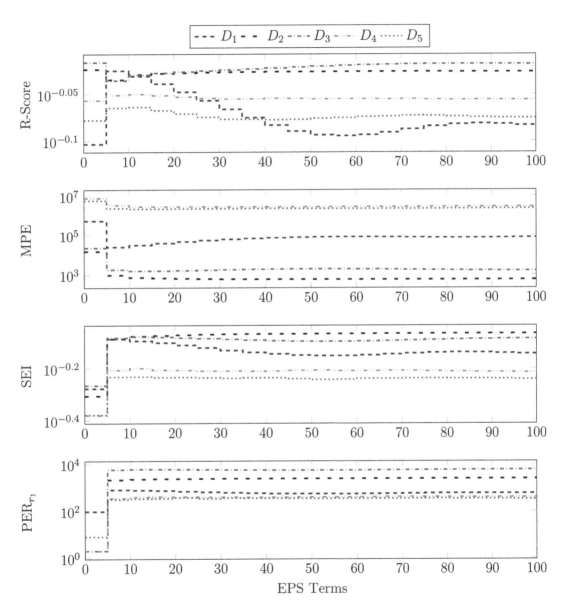

FIGURE 3.5 Error preventive ARMA forecast analysis for 20-minute prediction interval

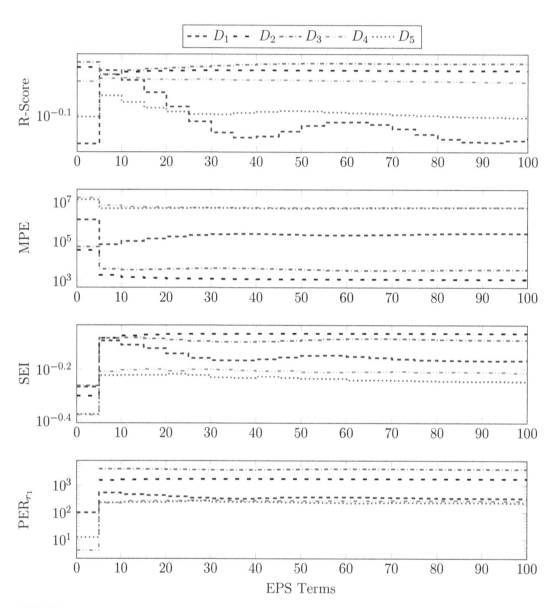

FIGURE 3.6 Error preventive ARMA forecast analysis for 30-minute prediction interval

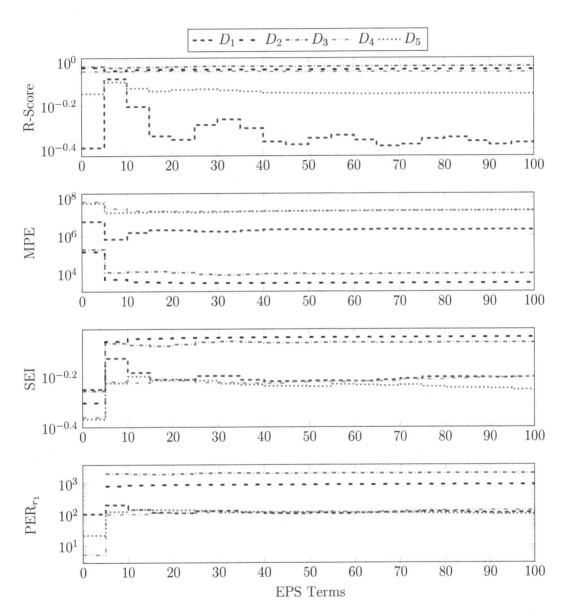

FIGURE 3.7 Error preventive ARMA forecast analysis for 60-minute prediction interval

The forecast accuracy measured using R-Score is shown in Fig. 3.8. It is evident that the value of R-Score lies between 0 and 1, and larger the value, better the accuracy. From results, it can be observed that the R-Score is improved after the inclusion of the error prevention score. The error preventive ARMA model witnesses an improvement between 21% and 183%. Similarly, the performance of both models is compared on SEI and the results are depicted in Fig. 3.9. Similar to R-Score, the value of the SEI metric ranges between 0 and 1, and a higher value indicates better model accuracy. The MPE-based forecast results are depicted in Fig. 3.10. As opposed to R-Score and SEI, the MPE metric advocates a model with lower values. The results clearly show the reduction in MPE by error preventive ARMA.

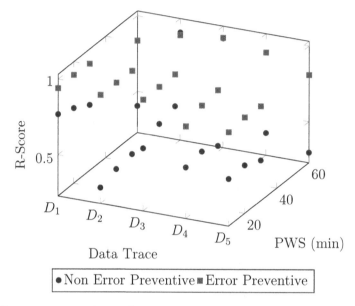

FIGURE 3.8 R-score comparison of non-error preventive and error preventive ARMA

A critical comparison of the error preventive and non-error preventive ARMA process shows that reasonable improvement is obtained through the error prevention scheme. For instance, $\mathcal{WP}^{\mathcal{EP}}_{\text{ARMA}}$ reduced the MPE at least by a factor of 72% over the non-error preventive counterpart. Figure 3.11 shows the forecast accuracy on PER which validates the better performance of the error preventive process. It is verified from the results that the $\mathcal{WP}^{\mathcal{EP}}_{\text{ARMA}}$ generates more forecasts with lower error magnitude. Further, the performance is assessed on MoP, and results are shown in Figs. 3.12 and 3.13 corresponding to the positive and negative magnitudes respectively. A model with a lower value of MoP^{-} is always preferred over a model having higher values and the error preventive ARMA has significantly reduced the count. Thus, a resource management framework enabled with an error preventive forecasting model helps in keeping reasonably enough computing resources active to fulfill user workload demands. In general, the error preventive model improves the forecast quality of different data sets measured using various metrics. However, it is on the cost of a little excess use of resources.

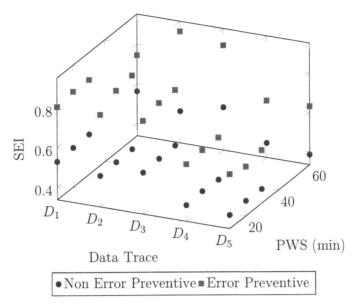

FIGURE 3.9 SEI comparison of non-error preventive and error preventive ARMA

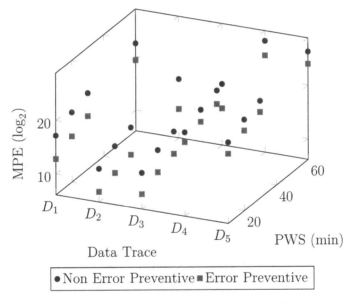

FIGURE 3.10 MPE comparison of non-error preventive and error preventive ARMA

3.4.2 Error Preventive Autoregressive Integrated Moving Average

The more general form of an ARMA model includes the integration of time series data and it is commonly referred to as ARIMA. The ARIMA process combines three operations viz. Autoregression (AR), integration (I), and moving average (MA). The ARIMA process is a preferable choice for non-stationary time series data as it is capable of transforming the non-stationary data into stationary data by the means of differentiation or integration. In general, an ARIMA process is denoted as

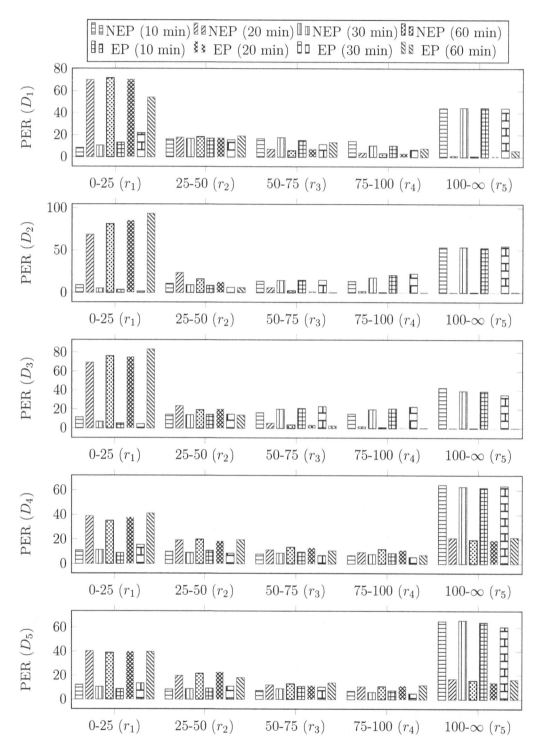

FIGURE 3.11 PER comparison of non-error preventive and error preventive ARMA

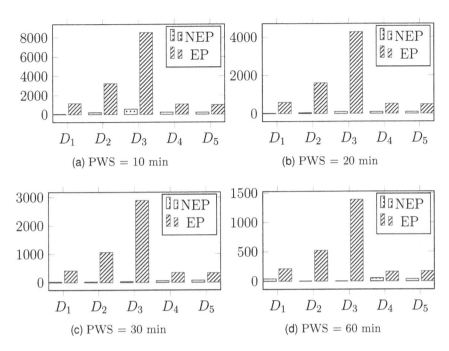

FIGURE 3.12 Positive magnitude comparison of non-error preventive and error preventive ARMA

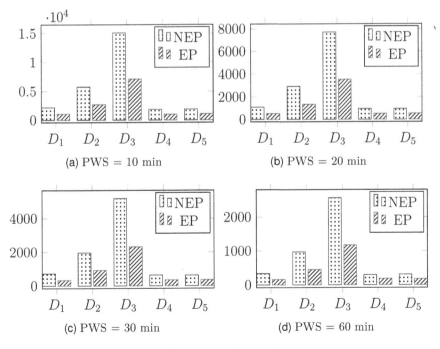

FIGURE 3.13 Negative magnitude comparison of non-error preventive and error preventive ARMA

ARIMA(p, d, q), where $p, d,$ and q represent the order of autoregression, integration, and moving average. The first order ARIMA(1,1,1) process-based $\mathcal{WP}_{\text{ARIMA}}$ can be written as shown in eq. (3.13), where the term $\tilde{x}_{t-1}$ represents the workload at time $t-1$ obtained after applying the first-order difference operator.

$$\hat{x}_t = \phi_1 \times \tilde{x}_{t-1} + \theta_1 \times \xi_{t-1} + \aleph_t \tag{3.13}$$

A prediction model based on the ARIMA process and enabled with error prevention is represented as $\mathcal{WP}_{\text{ARIMA}}^{\mathcal{EP}}$ and shown in eq. (3.14), where EPS_t denotes the error prevention score at time t. Again, a detailed experimental study is conducted to assess the performance of error preventive and non-error preventive ARIMA-based forecasting models. In an error preventive model, it is critically important to find out the reasonably suitable length of the error feedback window to harness the power of the error prevention scheme. Figures 3.14, 3.15, 3.16, and 3.17 depict the corresponding results for 10, 20, 30, and 60-minute prediction intervals respectively. It can be verified from the results that a lower value of feedback window length is a better choice which indicates that recent forecast errors are more useful in capturing the pattern. Therefore, the forecast accuracy of both models is compared on feedback window of length five terms. The R-score-based forecast accuracy is shown in Fig. 3.18 and it can be seen that a notable improvement is obtained by the error preventive model. The $\mathcal{WP}_{\text{ARIMA}}^{\mathcal{EP}}$ witnesses an improvement of up to 182% over $\mathcal{WP}_{\text{ARIMA}}$. Similarly, an improvement of up to 92% is obtained in SEI by $\mathcal{WP}_{\text{ARIMA}}^{\mathcal{EP}}$ as shown in Fig. 3.19. The forecast accuracy is measured on another standard parameter i.e. MPE and corresponding results are shown in Fig. 3.20. Again, the results show that the error prevention-based model improves the forecast quality.

$$\hat{x}_t = \phi_1 \times \tilde{x}_{t-1} + \theta_1 \times \xi_{t-1} + \aleph_t + EPS_t \tag{3.14}$$

The forecast quality is also measured on two self-proposed metrics i.e. PER and MoP. Figure 3.23 shows the results obtained on PER which clearly depict the significant increment in the number of data points falling in r_1. Subsequently, the error preventive model substantially reduces the number of forecasts from the high margin error ranges. Another criterion of quality measurement is MoP which checks the magnitude of forecast errors. As per the assumption that a cloud management system would prefer a forecasting application with more data points forecasted with positive error magnitude which is supported by various reasons. The corresponding results are depicted in Figs. 3.21 and 3.22 for the positive and negative magnitude of forecast errors. The error preventive model has significantly improved the number of forecasts with positive magnitude errors as opposed to the non-error preventive counterpart.

3.4.3 Error Preventive Exponential Smoothing

Exponential smoothing is another popular forecasting model based on time series analysis. This model suggests including both information i.e. actual and predicted workload values and can be represented as $\hat{x}_t = \alpha x_{t-1} + (1-\alpha)\hat{x}_{t-1}$ (see eq. (2.5)). It

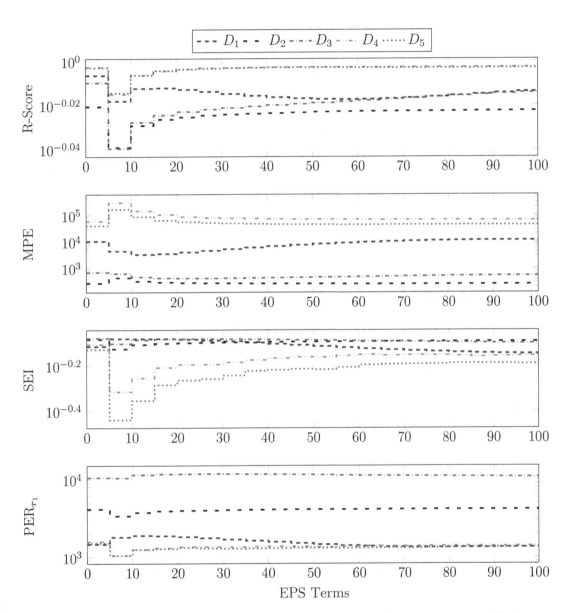

FIGURE 3.14 Error preventive ARIMA forecast analysis for 10-minute prediction interval

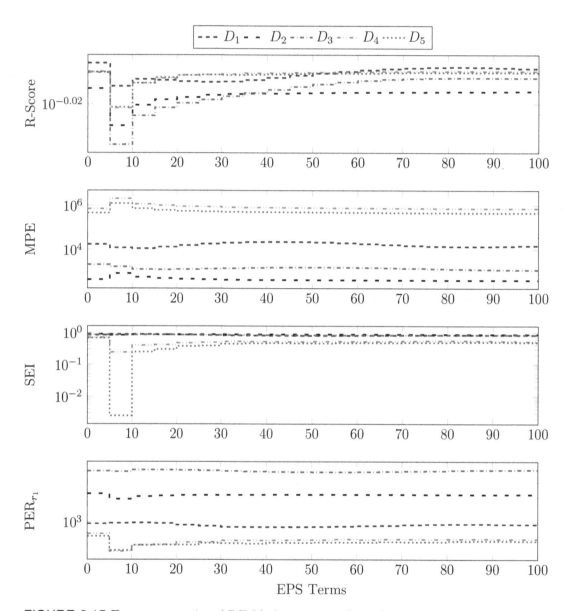

FIGURE 3.15 Error preventive ARIMA forecast analysis for 20-minute prediction interval

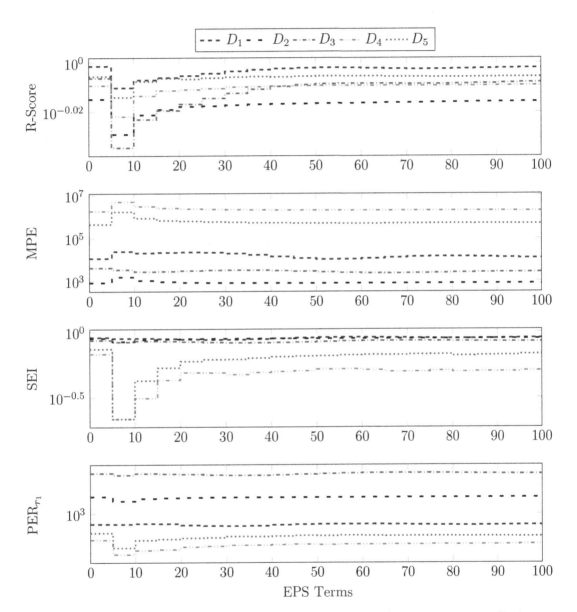

FIGURE 3.16 Error preventive ARIMA forecast analysis for 30-minute prediction interval

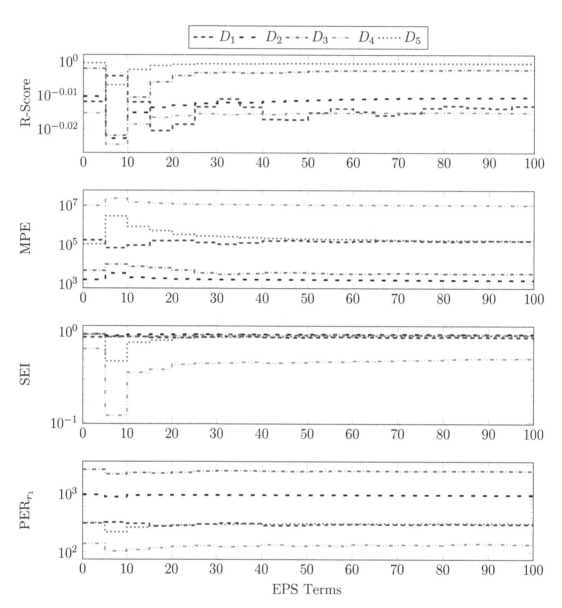

FIGURE 3.17 Error preventive ARIMA forecast analysis for 60-minute prediction interval

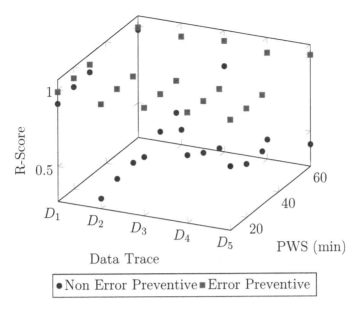

FIGURE 3.18 R-score comparison of non-error preventive and error preventive ARIMA

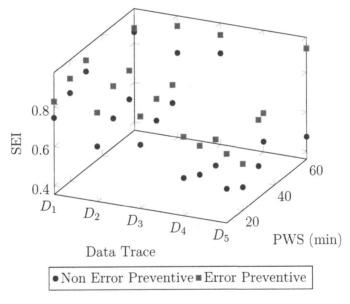

FIGURE 3.19 SEI comparison of non-error preventive and error preventive ARIMA

also argues that a more recent forecast may contribute more in modeling a time series data due to the factor of localization in time. The associated weight corresponding to the previous forecasts gradually gets reduced as the forecast becomes older in time. The error preventive forecasting model based on exponential smoothing ($\mathcal{WP}_{ES}^{\mathcal{EP}}$) is shown in eq. (3.15), where α is a constant term associated with smoothing which lies in (0, 1). The value of α determines the contribution weights associated with the previous workload values and the remaining weight is assigned to the recent predictions. If the actual workload at the previous time instance is stable, a lower

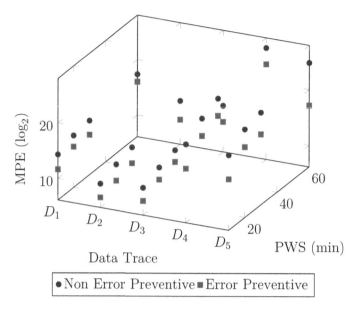

FIGURE 3.20 MPE comparison of non-error preventive and error preventive ARIMA

value of α is selected. Otherwise, the value of α reaches 1, if the previous workload value recorded any fluctuations. The value of α was selected based on an experimental analysis which records the performance of the model at different values of α and the best performing value had been selected for further experiments.

In this case also, a set of experiments are conducted to find the suitable length of the error feedback window. Again, the results recommended using a lower length of windows and it can be seen in Figs. 3.24, 3.25, 3.26, and 3.27 corresponding to prediction intervals of 10, 20, 30, and 60 minutes respectively. Therefore, the forecast results with a feedback window length of five error terms are reported. Figure 3.28 depicts the forecast accuracy measured on the R-Score metric and it can be seen that the error preventive model has significantly improved the quality of forecasts. Similarly, the non-error preventive model is outperformed on SEI as shown in Fig. 3.29. Another interesting point about SEI-based results is that the forecast accuracy improves as the prediction interval increases. Similarly, the error preventive model substantially improves the mean prediction error over $\mathcal{WP}_{ES}$. For instance, the $\mathcal{WP}_{ES}^{\mathcal{EP}}$ reduced the forecast error from 15% to 100% over its standard counterpart as shown in Fig. 3.30.

$$\hat{x}_t = \alpha x_{t-1} + (1 - \alpha)\hat{x}_{t-1} + EPS_t \qquad (3.15)$$

The forecast accuracy measured on two self introduced metrics is also reported. Figure 3.33 shows the results of prediction in error range metrics and a similar trend is observed where the error magnitude is relatively reduced with a significant factor. While Figs. 3.31 and 3.32 depict the performance of both models on the magnitude of the forecast. In this case as well, the error preventive model is performing better over a non-error preventive model.

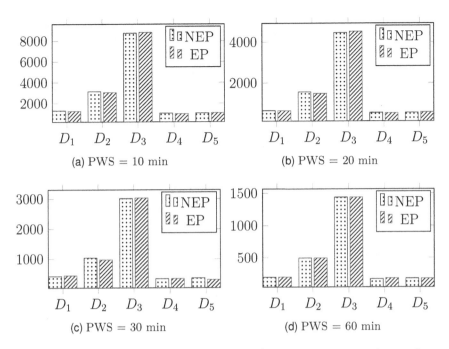

FIGURE 3.21 Positive magnitude comparison of non-error preventive and error preventive ARIMA

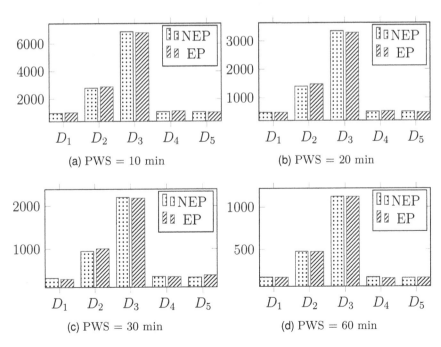

FIGURE 3.22 Negative magnitude comparison of non-error preventive and error preventive ARIMA

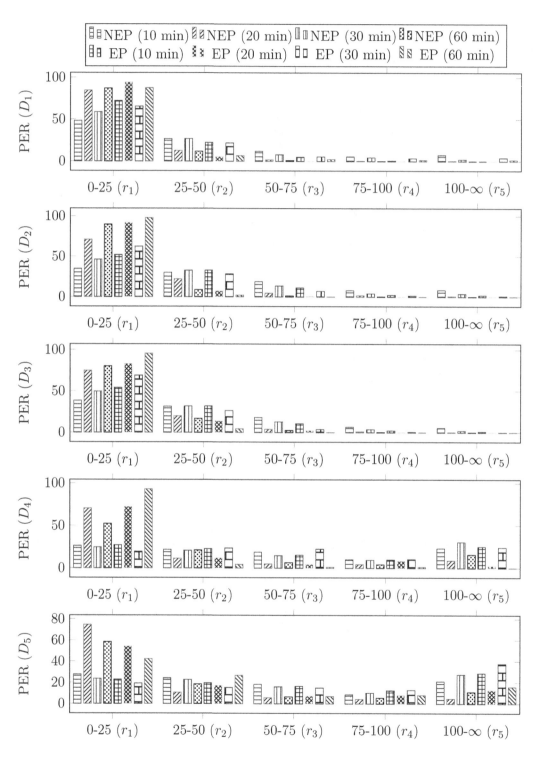

FIGURE 3.23 PER comparison of non-error preventive and error preventive ARIMA

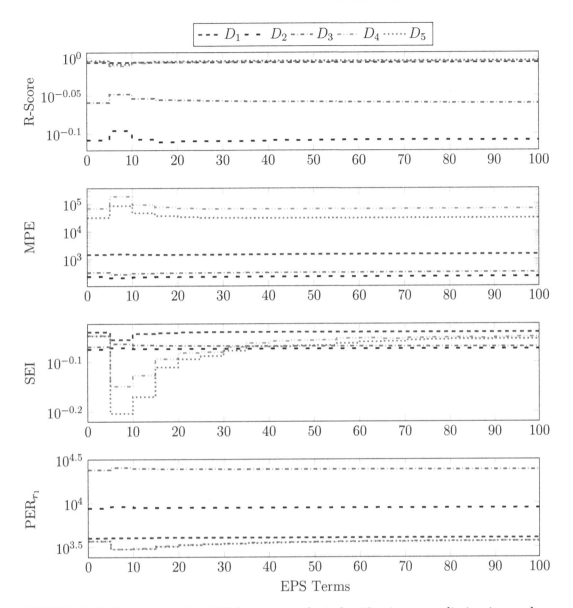

FIGURE 3.24 Error preventive ES forecast analysis for 10-minute prediction interval

3.5 PERFORMANCE EVALUATION

The experimental results presented in the above section show that the error preventive models are performing better than the non-error preventive models. In this section, detailed comparison and statistical analysis are conducted.

3.5.1 Comparative Analysis

According to the experimental findings presented in the above section, it is evident that the error prevention scheme is able to improve the performance of a forecasting model. For instance, $\mathcal{WP}^{\mathcal{EP}}_{\text{ARMA}}$ successfully improved the R-Score by a factor of

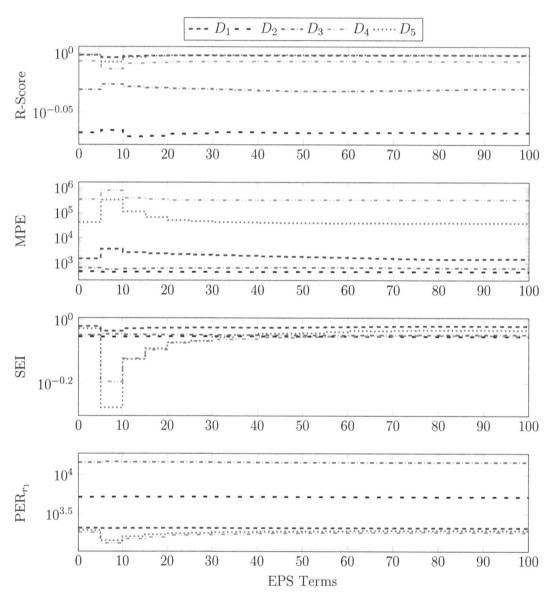

FIGURE 3.25 Error preventive ES forecast analysis for 20-minute prediction interval

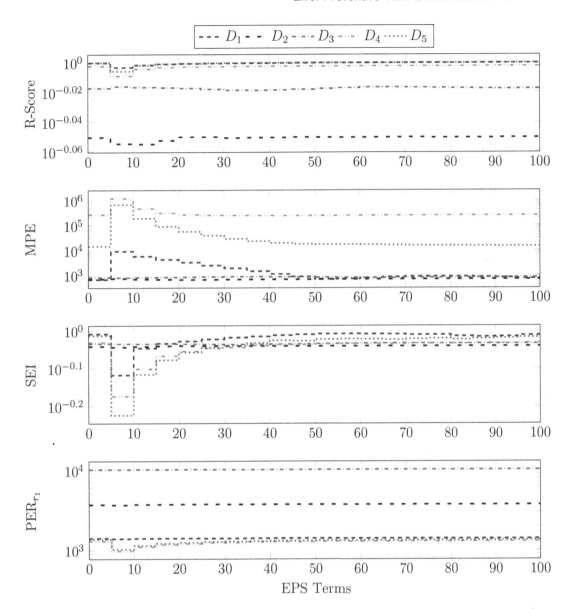

FIGURE 3.26 Error preventive ES forecast analysis for 30-minute prediction interval

22.4% while forecasting D_1 on a 10-minute prediction interval. The improvement term here refers to the relative improvement in the performance of a model over another model. A positive value of relative improvement in R-Score and SEI indicates better performance. The error preventive ARMA model achieves an improvement on R-Score up to 148.0%, 183.9%, 50.1%, 142.1%, and 173.8% for D_1, D_2, D_3, D_4, and D_5 respectively as shown in Fig. 3.34. Among the datasets used for the experimental study, D_2 forecasts are much better than others. The Calgary Trace received the best forecast quality on prediction intervals of size 10, 20, and 30 minutes. On the other hand, D_5 forecasts are best at 60-minute forecast intervals. Similarly, the ARIMA model was able to achieve a maximum relative improvement up to 10.2%, 182.3%,

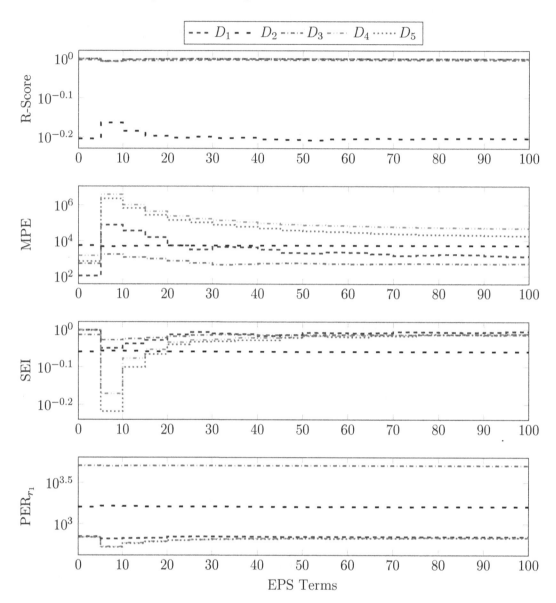

FIGURE 3.27 Error preventive ES forecast analysis for 60-minute prediction interval

49.0%, 142.6%, and 140.9% for D_1, D_2, D_3, D_4, and D_5 correspondingly. The Memory Trace witnesses the best improvement among the used data traces. A similar pattern was observed on comparing the performance of the exponential smoothing model with its error preventive counterpart approach. The next metric of interest is SEI and it was noticed that $WP_{ARMA}^{\mathcal{EP}}$ gets maximum improvement up to 53.4%, 80.6%, 53.7%, 52.9%, and 63.5% for D_1, D_2, D_3, D_4, and D_5 correspondingly as depicted in Fig. 3.35. Subsequently, the $WP_{ARIMA}^{\mathcal{EP}}$ model notices a relative performance improvement of up to 95.4%. As opposed to the R-Score and SEI, a negative value of relative performance comparison indicates the better performance measured using MPE. It is due to the fact that the least value of MPE indicates the best forecast. It can be seen that the error

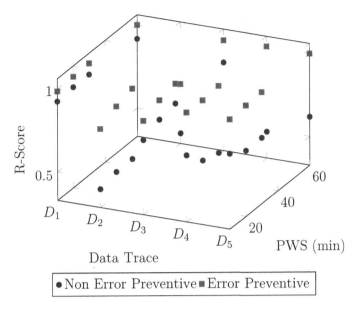

FIGURE 3.28 R-score comparison of non-error preventive and error preventive ES

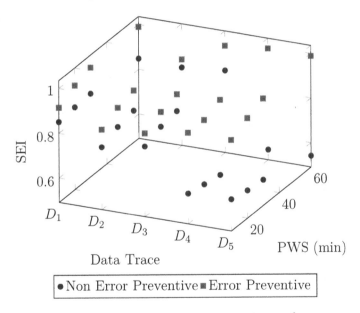

FIGURE 3.29 SEI comparison of non-error preventive and error preventive ES

preventive version of ARMA model reduces the forecast errors (MPE) up to 95.5%, 98.1%, 96.0%, 84.7%, and 79.5% for data traces D_1, D_2, D_3, D_4, and D_5 respectively as shown in Fig. 3.36. Whereas the $\mathcal{WP}^{\mathcal{EP}}_{\text{ARIMA}}$ and $\mathcal{WP}^{\mathcal{EP}}_{\text{ES}}$ subsequently reduce the forecast errors up to 99.5% and 100.0% respectively. Thus, the comparative analysis advises the inclusion of the error prevention scheme for better forecasts.

Further, the comparative analysis is conducted on the performance measured using the newly introduced performance indicators. It can be observed from the results that 70.00%, 71.72%, 70.62%, and 54.18% of D_1 predictions obtained from $\mathcal{WP}^{\mathcal{EP}}_{\text{ARMA}}$ on

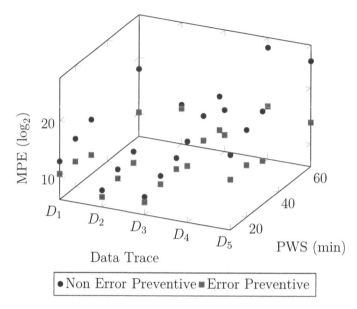

FIGURE 3.30 MPE comparison of non-error preventive and error preventive ES

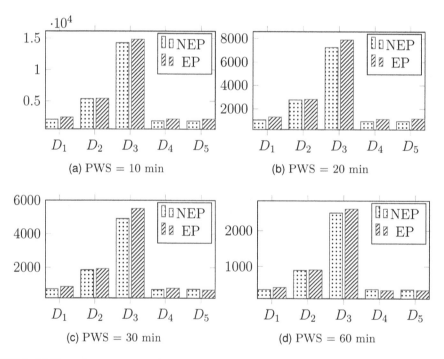

(a) PWS = 10 min

(b) PWS = 20 min

(c) PWS = 30 min

(d) PWS = 60 min

FIGURE 3.31 Positive magnitude comparison of non-error preventive and error preventive ES

prediction intervals of duration 10, 20, 30, and 60 minutes fall in PER_{r_1} as oppose to its non-error preventive counterpart model which could generate 8.34%, 10.58%, 13.06%, and 21.77% forecasts in PER_{r_1}. Similarly, it was found that $\mathcal{WP}^{\mathcal{EP}}_{\text{ARIMA}}$ generates 42.36% of forecasts in PER_{r_1} as opposed to the non-error preventive ARIMA model

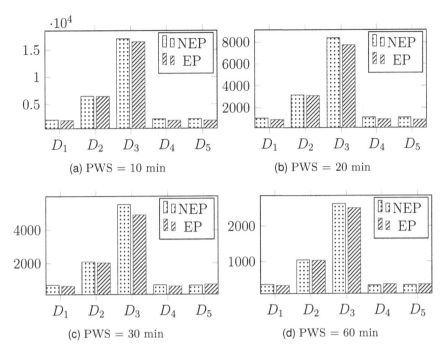

FIGURE 3.32 Negative magnitude comparison of non-error preventive and error preventive ES

that generates 19.25% of forecasts. The exponential smoothing also observed a similar trend where 76.85% and 27.09% forecasts of error preventive and non-error preventive models belong to PER_{r_1}. It shows a substantial improvement in terms of learning the pattern of actual workloads. However, the comparison of two different approaches based on a set of experiments does not signify the superiority or inferiority of one over the other. The statistical analysis helps in establishing the significance in the performance of two or more models.

3.5.2 Statistical Analysis

This study uses the Wilcoxon test [130] to find the answer to the question of performance superiority of the error preventive model over the non-error preventive model. The test follows a null hypothesis (H_0^{WC}) which assumes the equivalence between the performance of two approaches. The H_0^{WC} is not accepted if the test finds any significant difference between the performance of the approaches, otherwise, it accepts the null hypothesis. The corresponding results are listed in Table 3.1 which shows that the test does not accept the null hypothesis except in one case. The Wilcoxon test establishes a relationship that the error preventive model significantly improves the performance of their corresponding non-error preventive models.

The Wilcoxon test successfully reported the presence or absence of significant differences in the performance of the two models. But it is unable to report the model with best or worse performance. The Friedman test with Finner post-hoc analysis is capable of doing the same. Thus, an analysis obtained from the Friedman

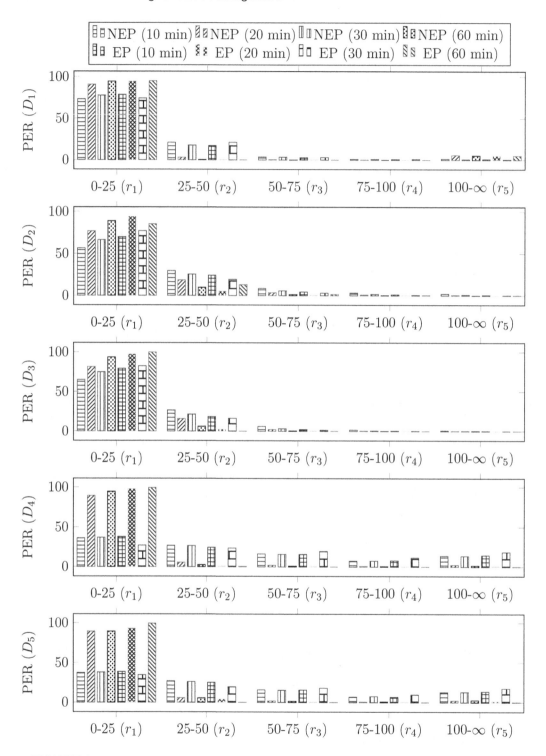

FIGURE 3.33 PER comparison of non-error preventive and error preventive ES

test with Finner post-hoc analysis is reported. First, the Friedman test compares the performance of multiple models around a null hypothesis (H_0^{FR}) which assumes

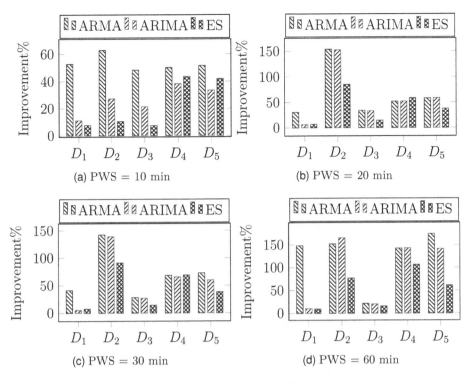

FIGURE 3.34 R-score performance relative improvement

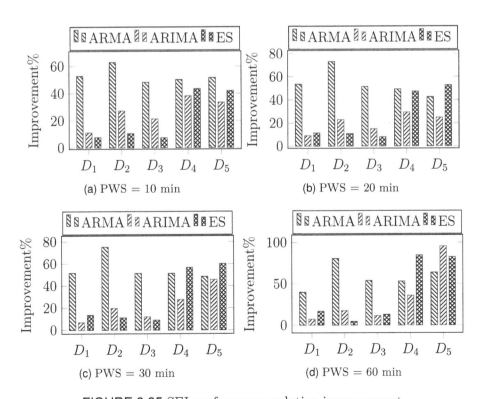

FIGURE 3.35 SEI performance relative improvement

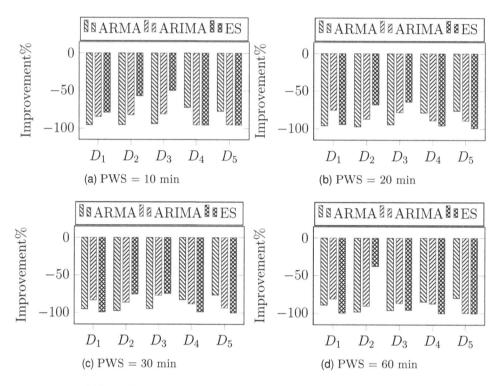

FIGURE 3.36 MPE performance relative improvement

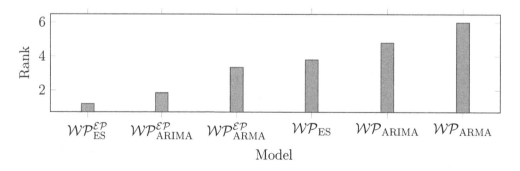

FIGURE 3.37 Friedman test ranks of non-error preventive and error preventive models

TABLE 3.1 Wilcoxon test statistics for error preventive and non-error preventive time series forecasting model

	Accuracy Metric	p-value	Result
$\mathcal{WP}_{ARMA}$ vs $\mathcal{WP}_{ARMA}^{\mathcal{EP}}$	R-Score	0.000089	H_0^{WC}.R
$\mathcal{WP}_{ARIMA}$ vs $\mathcal{WP}_{ARIMA}^{\mathcal{EP}}$	R-Score	0.000089	H_0^{WC}.R
$\mathcal{WP}_{ES}$ vs $\mathcal{WP}_{ES}^{\mathcal{EP}}$	R-Score	0.000088	H_0^{WC}.R
$\mathcal{WP}_{ARMA}$ vs $\mathcal{WP}_{ARMA}^{\mathcal{EP}}$	MPE	0.000089	H_0^{WC}.R
$\mathcal{WP}_{ARIMA}$ vs $\mathcal{WP}_{ARIMA}^{\mathcal{EP}}$	MPE	0.000089	H_0^{WC}.R
$\mathcal{WP}_{ES}$ vs $\mathcal{WP}_{ES}^{\mathcal{EP}}$	MPE	0.000088	H_0^{WC}.R
$\mathcal{WP}_{ARMA}$ vs $\mathcal{WP}_{ARMA}^{\mathcal{EP}}$	SEI	0.000088	H_0^{WC}.R
$\mathcal{WP}_{ARIMA}$ vs $\mathcal{WP}_{ARIMA}^{\mathcal{EP}}$	SEI	0.000089	H_0^{WC}.R
$\mathcal{WP}_{ES}$ vs $\mathcal{WP}_{ES}^{\mathcal{EP}}$	SEI	0.851925	H_0^{WC}.A
$\mathcal{WP}_{ARMA}$ vs $\mathcal{WP}_{ARMA}^{\mathcal{EP}}$	PER	0.000140	H_0^{WC}.R
$\mathcal{WP}_{ARIMA}$ vs $\mathcal{WP}_{ARIMA}^{\mathcal{EP}}$	PER	0.000089	H_0^{WC}.R
$\mathcal{WP}_{ES}$ vs $\mathcal{WP}_{ES}^{\mathcal{EP}}$	PER	0.000089	H_0^{WC}.R

TABLE 3.2 Finner test post-hoc analysis of error preventive and non-error preventive time series forecasting models

	$\mathcal{WP}_{ES}^{\mathcal{EP}}$ vs $\mathcal{WP}_{ARMA}$	$\mathcal{WP}_{ES}^{\mathcal{EP}}$ vs $\mathcal{WP}_{ARIMA}$	$\mathcal{WP}_{ES}^{\mathcal{EP}}$ vs $\mathcal{WP}_{ES}^{\mathcal{EP}}$	$\mathcal{WP}_{ES}^{\mathcal{EP}}$ vs $\mathcal{WP}_{ARMA}^{\mathcal{EP}}$	$\mathcal{WP}_{ES}^{\mathcal{EP}}$ vs $\mathcal{WP}_{ARIMA}^{\mathcal{EP}}$
Statistics	8.11348	6.08511	4.39480	3.63416	1.09870
Adjusted p-value	0.00000	0.00000	0.00002	0.00035	0.27190
Result	H_0^{FN}.R	H_0^{FN}.R	H_0^{FN}.R	H_0^{FN}.R	H_0^{FN}.A

the equivalence among the models' performance. The test does not accept the null hypothesis with 0.0 p-value and 212.70 statistic value. The detailed results are shown in Fig. 3.37 and it can be seen that the models equipped with error prevention scheme obtained better ranks. The error preventive ES model achieves the best rank among all models. The Finner test results presented in Table 3.2 validate the claims of the Wilcoxon test as it also does not find any significant difference between the performance of error preventive and non-error preventive exponential smoothing models. Thus, the experimental and statistical observations advocate the use of error preventive methods over non-error preventive methods for better forecasts.

CHAPTER 4

Metaheuristic Optimization Algorithms

T HE real-world problems such as routing, resource allocation, and engineering designs are multimodal and depict a non-linear behavior. These problems are generally modeled as a constraint optimization problem with one or more decision problems and most of these problems belong to the NP-hard class [16, 86]. The different optimization methods are being used and explored to solve such problems. In general, trajectory-based algorithms and population-based algorithms are two different categories of optimization algorithms. An algorithm that explores the solution space using a single solution is referred to as a trajectory-based algorithm. Whereas an algorithm that uses a set of solutions to find the optimal solution is called a population-based algorithm. This chapter focuses on a variety of widely used population-based algorithms (see Fig. 4.1) and their performance in predicting the workloads in cloud environment [5,7−9,20,39,65,95,103].

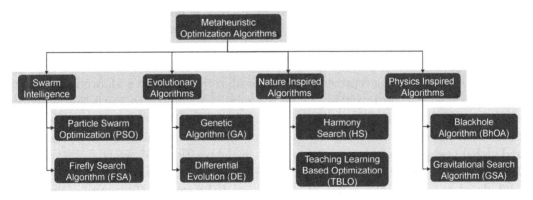

FIGURE 4.1 Population-based metaheuristic optimization algorithms' taxonomy

4.1   SWARM INTELLIGENCE ALGORITHMS IN PREDICTIVE MODEL

A swarm algorithm follows the principles of cooperative behavior of a group of homogeneous agents in nature such as birds, fish, antss etc. These algorithms are generic and can be applied to every optimization problem. In this category, particle swarm optimization and firefly search algorithms are two widely used algorithms. In subsequent subsections, these algorithms are discussed and their performance is assessed in detail.

4.1.1 Particle Swarm Optimization

The particle swarm optimization commonly referred to as PSO was introduced by J. Kennedy and R. Eberhart in 1995 [68]. The algorithm follows the principles of the social behavior of birds i.e. each member of the group searches for a better food source in the direction of its leader. In this study, the algorithm is used to find the suitable combination of synaptic connection weights for the neural network-based predictive framework. Let N be the number of particles in the search space, where each particle denotes a neural network-based predictive model composed with different combinations of weights. In PSO, each individual or particle has two components viz. position and velocity corresponding to every dimension of search space. Our experimental setup randomly initiates the positions and velocities as given in eqs. (4.1) and (4.2), where upper and lower bounds are represented as $ub_j = +1$ and $lb_j = -1$ respectively, r represents a random real number in $(0, 1)$. Similarly, $s_i.p_j$ and $s_i.v_j$ correspond to the position and velocity of ith individual in j^{th} dimension.

$$s_i.p_j = (ub_j - lb_j) \times r + lb_j \tag{4.1}$$

$$s_i.v_j = (ub_j - lb_j) \times r + lb_j \tag{4.2}$$

$$f_{cost} = \min_{\hat{x}_t} \left(\sqrt[2]{\frac{1}{m} \sum_{t=1}^{m} (x_t - \hat{x}_t)^2} \right) \tag{4.3}$$

After initialization of the population, each particle is evaluated on the training data and the corresponding fitness value is assigned which is measured using root mean squared error. The objective of the algorithm is to estimate the accurate information of upcoming workloads or close to actual workload i.e. minimizing the error in the forecasts, a corresponding representation of the objective function is shown in eq. (4.3). The particle swarm optimization helps the predictive model to learn the network weights in an iterative fashion. In each iteration, the particles move in the search space to look for a better fitness value by modifying their position and velocity. The local and global best particles impacts the updates in the position and velocity of a particle. The global best (*gbest*) particle has the best prediction accuracy among the swarm while local best is defined as per the topology of the swarm. This study considers that the personal best (*pbest*) of a particle is the local best. In some other cases, the swarm may have small groups and the best among the members of a group can be considered as the local best. The update procedures for the position and the velocity are shown in eqs. (4.4) and (4.5) respectively. Once the updates have occurred across the swarm, the *gbest* and *pbest* are updated accordingly. This process is repeated until termination criteria are met i.e. fixed number of iterations (250) in this study.

$$s_i.v(t+1) = \omega \times s_i.v(t) + c_1 \times r \times (s_i.p_{pbest} - s_i.p(t)) + c_2 \times r \times (s.p_{gbest} - s_i.p(t)) \tag{4.4}$$

$$s_i.p(t+1) = s_i.p(t) + s_i.v(t) \tag{4.5}$$

4.1.2 Firefly Search Algorithm

The Firefly Search Algorithm (FSA) is another widely used swarm intelligent algorithm that was introduced by Xin-She Yang [135, 136]. It is inspired by the flashing pattern and behavior of fireflies in a group that regularly produces the flash. A firefly is attracted towards another firefly which generates the flash with more intensity. The key components of the algorithm are flash intensity and attractiveness. The brightness of a firefly for another firefly depends on the distance between two and commonly brighter firefly attracts the other fireflies. The communication among the fireflies happens with an assumption that the population members are unisex and a firefly can attract any other member in the population. In this algorithm, the brightness intensity of the firefly indicates its fitness i.e. value observed using an objective function.

Firefly in the population is attractive in accordance with the intensity of the flash visible to its neighbors. Let β_0 be the attractiveness of an individual population member at $d = 0$, its attractiveness at distance d can be obtained from eq. (4.6). On the other hand, the movement of fireflies can be obtained from eq. (4.7), where the attraction is represented by the second term and the effect of randomization involved in the process is represented by the third term in the model (α_t is randomization parameter, ϵ_i^t is a vector of random numbers).

$$\beta = \beta_0 \times e^{-\gamma d^2} \tag{4.6}$$

$$s_i(t+1) = s_i(t) + \beta_0 \times e^{-\gamma d_{i,j}^2}(s_j(t) - s_i(t)) + \alpha_t \epsilon_i^t \tag{4.7}$$

In this chapter, the performance of the predictive models is assessed on three different values (5, 30, and 60 minutes) of the prediction window. The accuracy of the forecasts is measured using root mean squared error, and mean absolute error and results are depicted in Fig. 4.2. It can be seen that the FSA outperformed the PSO in more number of experiments. To be more specific, according to the RMSE results, FSA outperformed PSO in nine instances and achieves equal accuracy (measured up to three decimal digits) in four instances. The particle swarm optimization algorithm was able to get better results in two experiments only. Similar trends are observed in the results based on MAE also. An initial analysis of the results supports the superiority of FSA over PSO. These findings are assessed using Wilcoxon signed-rank test to find the statistical significance of the results. It computes positive (R_{WC}^+) and negative (R_{WC}^-) ranks based on the difference between the results of the two approaches. If the resultant value of accuracy obtained from FSA is lower, it would update the negative rank of FSA (since the problem under consideration is a minimization problem, the higher negative rank would be preferable); if the resultant value of accuracy

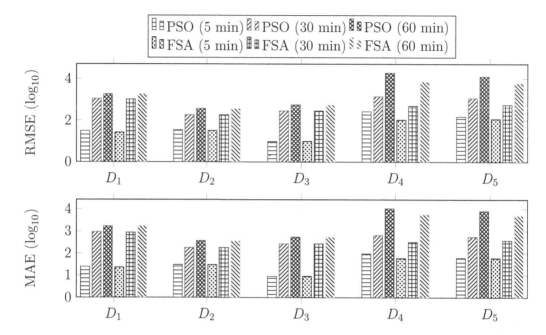

FIGURE 4.2 Forecast accuracy comparison of swarm intelligence based prediction models

obtained from FSA is higher, the negative rank of FSA will be updated. When the performance of both approaches is the same, both ranks R_{WC}^- and R_{WC}^+ get updated. The corresponding results are shown in Fig. 4.3 and it can be seen that the FSA gets a higher negative rank for three different data sets. Moreover, the test was failed in rejecting the null hypothesis (H_0^{WC}) for D_1, D_2, and D_3 while it successfully rejected the null hypothesis for the remaining two data traces i.e. D_4 and D_5 which indicates the superiority of FSA based predictive model.

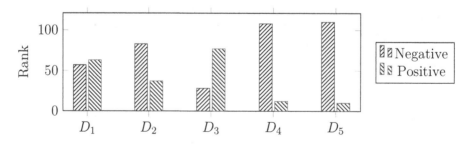

FIGURE 4.3 Wilcoxon test statistics of swarm intelligence based prediction models

4.2 EVOLUTIONARY ALGORITHMS IN PREDICTIVE MODEL

The algorithms that adapt the principles from evolution mechanisms such as evolution in biology fall into this category. Two widely used algorithms are considered for the purpose of studying their performance on cloud workload forecasting.

4.2.1 Genetic Algorithm

The genetic algorithm (GA) is one of the most popular and widely used algorithms in this family. The algorithm opts the principles from Darwin's evolution theory and uses them in the computing world. The algorithm was proposed by J. Holland in 1960s [58]. The algorithm borrows the terminology from biological evolution. For instance, every individual in the solution space is referred to as a chromosome, which is a collection of genes. Similar to other population-based optimization algorithms, GA also assigns the fitness to each of the population members after assessing their performance on training data. An iterative process is executed after fitness assignment which generates offsprings using crossover and mutation operators followed by survivor selection [50].

This study uses the Roulette wheel selection to select two parents (s_{k_1} and s_{k_2}) that who participate in the process of reproduction. The roulette wheel selection is a fair selection method as it allows an individual to get selected as per their fitness value which means a solution having better fitness value has more chances of selection. Further, it applies the single-point crossover operator to generate two offsprings from the parents. The parents are split into two parts from a randomly selected point and their tails get exchanged with the probability of CR (see eq. (4.8)). After crossover is performed, the newly generated solutions or offsprings go through another reproduction operation called a mutation. The mutation operator reinitialize randomly selected one of the values in the solution space as given in eq. (4.9), where R denotes the randomly selected dimension or position and r represents the reinitialized number in the solution space. The mutation operation explores the search space and avoids getting stuck in local optima very easily. Also, it helps in keeping up the population diversity, and, thus premature convergence is prevented. Afterward, the new solutions generated from the process of reproduction are assessed on the objective function and a fitness value is assigned to each of them. The algorithm selects N solutions among the old and new populations on the basis of individual's fitness. The study employs the survival of the best mechanism to select N solutions to participate in the next iteration. The entire process is repeated until the termination criteria are met or approximated solution is achieved, whichever is earlier.

$$u_{k_1,j} = \begin{cases} s_{k_1,j}, & \text{if } CR \leq r \\ s_{k_2,j}, & \text{otherwise} \end{cases} \tag{4.8}$$

$$v_{i,j} = \begin{cases} r, & \text{if } j == R \\ u_{i_2,j}, & \text{otherwise} \end{cases} \tag{4.9}$$

4.2.2 Differential Evolution

Differential evolution (DE) is a numerical optimizer developed by R. Storn and K. V. Price [105, 126]. This algorithm uses the concepts of vector manipulation and explores the search space to find the optimal solution. A population of randomly generated solutions is initialized followed by the assessment of each member on the objective function and corresponding fitness assignment. This algorithm varies with the other algorithms on the usage of reproduction operators. For instance, first it selects the base solution or vector s_i and three distinct random solutions $(s_{r_1}, s_{r_2}, s_{r_3})$ such that $i \neq r_1 \neq r_2 \neq r_3$. The difference between the two randomly selected vectors is weighted by the factor of F and added to the third vector. This process is referred to as mutation and the newly generated solution is the mutant solution. To be specific, the above-mentioned operator is named as $DE/rand/1$ (refer eq. (4.10)) [64]. Afterward, the base and mutant solutions participate in the crossover operation to generate an offspring solution (u_i). The crossover operator selects the gene either from base or mutant solution with the probability of CR, at least one of the genes is selected from the mutant solution to ensure the information exchange from the mutant vector as shown in eq. (4.11), where r and R are random numbers in $(0, 1)$ and $[1, D]$ respectively. The offspring solutions are assessed on the objective function and their fitness value is computed. Similar to the other algorithms, the population for the next iteration is selected using survival of the fittest.

$$v_i = s_{r_1} + F \times (s_{r_2} - s_{r_3}) \tag{4.10}$$

$$u_{i,j} = \begin{cases} v_{i,j}, & \text{if } (r \leq CR \lor j == R) \\ s_{i,j}, & \text{otherwise} \end{cases} \tag{4.11}$$

A set of experiments were conducted to evaluate the performance of both algorithms. The same experimental settings are used. Figure 4.4 shows the performance of the algorithms on different combinations of parameters including prediction windows size and data traces. A trend of similarity is observed in the performance. However, the performance of the GA was slightly better over DE in half of the experiments as per the MAE. As opposed to the MAE-based performance, the DE generates forecasts with lower RMSE. Thus, it is very difficult to say which algorithm is better. The statistical test also accepts the null hypothesis H_0^{WC} with significance levels $\aleph = 0.05$ and 0.1 for all data traces, thus, none of the algorithms can be said better. The Wilcoxon signed-ranks (depicted in Fig. 4.5) also support the statement that the both algorithms performed with similar accuracy except the fact that the GA was better in forecasting Saskatchewan Trace.

4.3 NATURE INSPIRED ALGORITHMS IN PREDICTIVE MODEL

In this section, we will discuss two widely used population-based search algorithms that follow the principles from nature. The algorithms are namely Harmony Search (HS) and Teaching Learning Based Optimization (TLBO) that use the concepts from human behavior.

4.3.1 Harmony Search

The harmony search algorithm is inspired by the Jazz musicians' improvisation in their music. It was first developed and proposed by Z. W. Geem et al. in 2001 [45] and has a range of applications. It follows the principle of improving the variations of individual musicians and coming up with a masterpiece of music. The algorithm

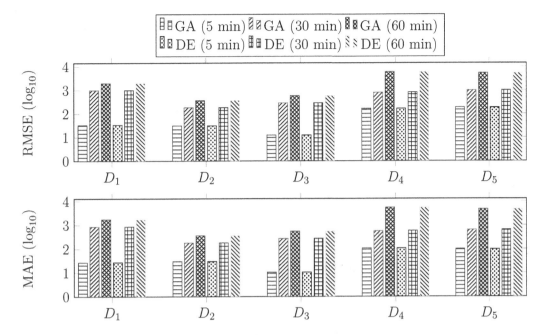

FIGURE 4.4 Forecast accuracy comparison of evolutionary algorithms based prediction models

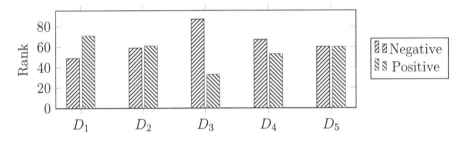

FIGURE 4.5 Wilcoxon test statistics of evolutionary algorithms based prediction models

encodes a solution in the search space as harmony among the musicians. The strength of the band i.e. the number of musicians in the band is used to encode the dimensions of the search problem, thus, each component of the solution represents the music composed by one distinct musician. The bounds on the decision variables are encoded as the pitch and range of an instrument being played by the respective musician. Similar to other algorithms, the population members are assessed and a corresponding fitness value is assigned to each member which is the appreciation received from the audience against the composition of the music. In order to get better appreciation, each musician tries to improve the music, thus, the quality of the overall solution improves. Each musician keeps the old music for the next round with the probability of $\mathcal{H}$ and regenerates the music with the probability of 1-$\mathcal{H}$ as shown in eq. (4.12). The pitch of newly generated harmony is adjusted as depicted in eq. (4.13). The change in the music is observed by δ which is computed as $\delta = FW \times randn()$, where FW indicates the maximum width of the allowed changes and $randn()$ is a function that generates a random number in $(0, 1)$. The newly generated solutions are assessed on the objective functions and the better solutions are kept for the next iteration.

$$u_{i,j} = \begin{cases} s_{i,j}, & \text{if } r \leq \mathcal{H} \\ rand(lb, ub), & \text{otherwise} \end{cases} \tag{4.12}$$

$$u_{i,j} = \begin{cases} u_{i,j} + \delta, & \text{if } r \leq \mathcal{P} \\ u_{i,j}, & \text{otherwise} \end{cases} \tag{4.13}$$

4.3.2 Teaching Learning Based Optimization

The teaching learning-based optimization algorithm opts the principles of the learning mechanism. It tries to model the effect of teaching quality on the learning process of a set of students [110]. It is a common understanding that the learning of a student is directly affected by the quality of the teacher's knowledge i.e. the students taught by a good teacher have higher chances of producing a good outcome as opposed to the set of students being taught by a teacher having less knowledge. The algorithm applies that concept in optimization as it designates the best solution in the population as the teacher and every other population member is treated as a student.

In human life, the common agenda of a teacher is to improve the knowledge of each student. Considering the fact that each individual has a different level of knowledge, the mean of the group appears to be a good representation of the knowledge of the population. Let μ_i and τ_i be the knowledge levels of the population and the teacher during i^{th} iteration respectively. As per the concept from the teaching-learning process in human life, the algorithm assigns a task to the teacher which is to bring the knowledge level of the population as close as possible to his own level of knowledge. In other words, each student in the population is allowed to use the knowledge of the teacher and other members of the population to improve their own knowledge. In the first phase of the learning i.e. teaching phase, the knowledge of a student is updated as shown in eq. (4.14), where u_i represents the student with updated knowledge. The $\Delta\mu_i$ is computed as $r(s_{best} - TF \times \mu_i)$, where s_{best} is the student with best learning, TF is the teaching factor, r is a random number in $(0, 1)$.

$$u_i = s_i + \Delta\mu_i \tag{4.14}$$

In the second phase of learning the student is allowed to learn by the means of interaction with other students. Let u_i and u_j be the randomly selected distinct students and they interact with each other. The knowledge of them is updated as depicted in eq. (4.15). It can be seen that in either of the mentioned cases, only weaker student learns, where f_{u_i} and f_{u_j} are the knowledge levels or fitness of respective students or solutions. If the knowledge of u_i is better than the knowledge of s_i, the change in the knowledge is accepted i.e. the updated solution is accepted for the next iteration otherwise s_i is continued in the process.

$$u_i = u_i + r \times (u_i - u_j) \qquad \text{if } f_{u_i} \leq f_{u_j}$$
$$u_j = u_j + r \times (u_j - u_i) \qquad \text{if } f_{u_j} \leq f_{u_i} \qquad (4.15)$$

The performance of HS and TLBO is assessed on the same set of experiments within the same experimental setup. Figure 4.6 shows the performance measure using RMSE and MAE. It can be noticed that the performance of TLBO is better than HS for 5 minutes forecast. Whereas for two other prediction intervals, both algorithms generate the forecasts with similar accuracy. It is observed that the TLBO outperforms HS in 33.33%, 33.33%, 40%, and 20% experimental instances based on RMSE and MAE values during network training while HS produced better results in 26.66%, 20%, 20%, and 0% experiments based on respective error metrics. Also, the Wilcoxon test accepted the null hypothesis for different values of significance level i.e. 0.05 and 0.1 both (Fig. 4.7). Thus, we did not find any significant difference in the performance of both algorithms.

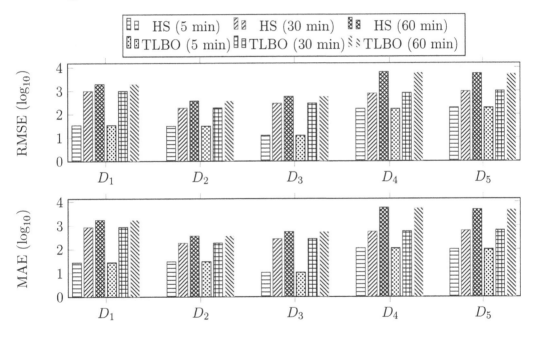

FIGURE 4.6 Forecast accuracy comparison of nature-inspired algorithms based prediction models

4.4 PHYSICS INSPIRED ALGORITHMS IN PREDICTIVE MODEL

The laws from physics have also been borrowed to develop population-based optimization algorithms. This study considers two of them which are widely used in different applications. The selected algorithms are gravitational search algorithm (GSA) and blackhole algorithm (BhOA).

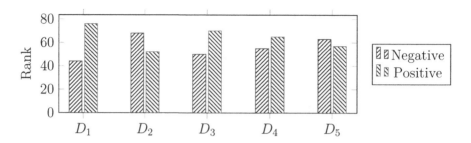

FIGURE 4.7 Wilcoxon test statistics of nature-inspired algorithms based prediction models

4.4.1 Gravitational Search Algorithm

Rashedi et al. proposed an optimization algorithm in 2009 which uses the concepts derived from the law of gravity and interaction between masses [111]. Based on the principle idea of the search process, the inventors named the algorithm as gravitational search algorithm. Similar to other population-based search algorithms, it also begins with a set of randomly initialized solutions in the search space. Each population member is assumed to be a mass in the environment which is governed by the laws of gravity and motion. According to gravity law, every mass in the environment attracts every other mass towards itself by applying a force which is called gravitational force. Whereas the law of motion governs the velocity of a mass which can be computed by taking change in the velocity and its fractional past velocity into consideration.

$$\mathcal{GF}_{ij}^{k}(t) = \mathcal{GC}(t) \times \frac{\mathcal{MS}_{i}^{p}(t) \times \mathcal{MS}_{j}^{a}(t)}{||s_i(t), s_j(t)||_{2+\epsilon}} \times (s_j^k(t) - s_i^k(t)) \qquad (4.16)$$

$$\mathcal{GF}_i^k(t) = \sum_{j=1, j \neq i}^{N} r \times \mathcal{GF}_{ij}^k(t) \qquad (4.17)$$

$$\alpha_i^k(t) = \frac{\mathcal{GF}_i^k(t)}{\mathcal{MS}_{ii}(t)} \qquad (4.18)$$

$$s_i.p_k(t+1) = s_i.p_k(t) + s_i.v_k(t+1) \qquad (4.19)$$

$$s_i.v_k(t+1) = r \times s_i.v_k(t) + \alpha_i^k(t) \qquad (4.20)$$

$$m_i(t) = \frac{f_{s_i}(t) - max(f_{s_j}(t))}{min(f_{s_j}(t)) - max(f_{s_j}(t))} \quad \forall j \qquad (4.21)$$

$$\mathcal{MS}_i(t) = \frac{m_i(t)}{\sum_{j=1}^{N} m_j(t)} \qquad (4.22)$$

Let $s_i.p_k$ and $s_i.v_k$ be the position and velocity in k^{th} dimension of ith solution. The force applicable on i^{th} mass from j^{th} mass in k^{th} dimension at time instance t can be modeled using eq. (4.16), where $\mathcal{MS}_i^p$ and $\mathcal{MS}_j^a$ represent the passive and active gravitational mass corresponding to the solution i and j respectively, $\mathcal{GC}(t)$ represents gravitational constant at time t which is a function of its initial value and time step t i.e. $\mathcal{GC}(t) = G(\mathcal{GC}(0), t)$, $||s_i(t), s_j(t)||_2$ represents the Euclidean distance between s_i and s_j at time t, and ϵ represents the small constant value. The total force applied on s_i in k^{th} dimension is obtained by summing up the randomly weighted forces applied by every other mass in the environment as shown in eq. (4.17), where r is a random number in $(0, 1)$. Thus, the acceleration of s_i in the direction k and at time t can be obtained from eq. (4.18), where $\mathcal{MS}_{ii}$ represents the inertia mass of s_i. Furthermore, the position and velocity of s_i get updated using eqs. (4.19) and 4.20 respectively, where r is a uniform random number in $[0, 1]$. In order to get the updated velocity, the fractional velocity of a solution is added to its acceleration. Whereas the updated position of a solution is observed by adding the updated velocity and its current position. The calculation of masses (gravitational and inertia) involves the fitness assessment as shown in eqs. (4.21) and (4.22), where $f_{s_i}(t)$ represents the fitness value of s_i at time t. This process gets repeated until termination criteria are met or the desired solution is achieved, whichever is earlier.

4.4.2 Blackhole Algorithm

The next algorithm that uses the principles derived from physics is the blackhole algorithm. It was developed and proposed by A. Hatamlou in 2013 [57] which is relatively new but has received good attention from the research community. As its name suggests, the algorithm derives the optimization process from the concepts of a blackhole, and being a population-based search algorithm, it uses a set of random solutions to explore the search space for an optimal solution. Every solution (s_i) in the search space is referred to as a star and the solution with the best fitness is referred to as a blackhole ($\mathcal{B}$). The search process is governed by the laws proposed in blackhole theory i.e. every star other than blackhole moves towards the blackhole due to the force applied on them. Thus, the updated position of a star can be modeled by adding its current position to the randomly weighted difference of $\mathcal{B}$ and the solution itself as shown in eq. (4.23), where r is a random number in $[0, 1]$ which is added to introduce the randomized behavior in the search process. Thus, every solution explores the search space to find a better solution with the guidance of the best solution achieved so far.

$$s_i(t+1) = s_i(t) + r \times (\mathcal{B} - s_i(t)) \tag{4.23}$$

The blackhole has a parameter associated with itself which is referred to as event horizon radius (ρ) and it is measured using eq. (4.24). If any star crosses this radius then the star enters into the blackhole and never comes back. Thus the star reaching into the event horizon radius of a blackhole gets collapsed and a new random star is generated to keep the population size uniform across the simulations. In order to check, if the star enters into the blackhole or not, its distance (d_i) from the blackhole is computed using eq. (4.25), where $f_\mathcal{B}$ is the fitness of blackhole and f_{s_i} is the fitness of star i. If the distance of a star from the blackhole is less than the radius of the blackhole, the star enters into the event horizon area of the blackhole. Furthermore, if a star finds a better solution in the search space than the existing blackhole, the blackhole is replaced with the newly discovered solution. The advantage of the BhOA over other population-based approaches are that it does not have any parameter to tune such as crossover rate, mutation rate, and others.

$$\rho = \frac{f_{\mathcal{B}}}{\sum_{i=1}^{N} f_{s_i}} \tag{4.24}$$

$$d_i = f_{\mathcal{B}} - f_{s_i} \tag{4.25}$$

The performance of GSA and BhOA is compared on the same set of experiments used in previous comparisons. Figure 4.8 shows the observed forecast accuracy of both algorithms on different data traces. It was observed that the predictive framework equipped with BhOA generates a better forecast in most of the cases. For instance, the D_3, D_4, and D_5 are better modeled by the BhOA. Whereas on other traces the GSA's performance was better in some of the experiments while BhOA's performance was better in some other cases. As per the experimental setup, the performance results are assessed using a statistical evaluation and the corresponding rankings are given in Fig. 4.9. As per the rankings, the performance of BhOA is better. The observations revealed by the rankings are also supported by the test as it rejected the null hypothesis with $\aleph = 0.05$ for D_2 and D_3, and accepts the H_0^{WC} in the rest of the cases. Thus, the BhOA can be said better as it significantly improves the performance at least on some data traces.

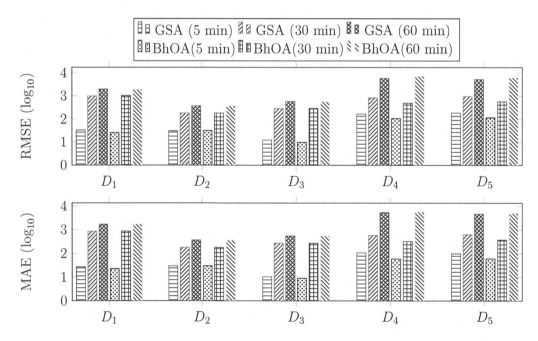

FIGURE 4.8 Forecast accuracy comparison of physics-inspired algorithms based prediction models

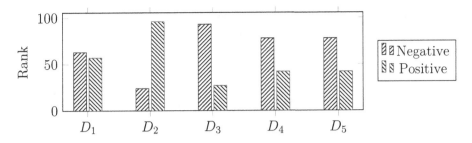

FIGURE 4.9 Wilcoxon test statistics of physics-inspired algorithms based prediction models

TABLE 4.1 Friedman test statistics of metaheuristic algorithms based prediction models

	D$_1$	D$_2$	D$_3$	D$_4$	D$_5$
χ^2	7.93333	23.54137	29.69355	34.47149	24.68835
p-value	0.33851	0.00137	0.00011	0.00001	0.00086
$\aleph = 0.05$	$H_0^{\mathrm{FR}}.\mathrm{A}$	$H_0^{\mathrm{FR}}.\mathrm{R}$	$H_0^{\mathrm{FR}}.\mathrm{R}$	$H_0^{\mathrm{FR}}.\mathrm{R}$	$H_0^{\mathrm{FR}}.\mathrm{R}$
$\aleph = 0.1$	$H_0^{\mathrm{FR}}.\mathrm{A}$	$H_0^{\mathrm{FR}}.\mathrm{R}$	$H_0^{\mathrm{FR}}.\mathrm{R}$	$H_0^{\mathrm{FR}}.\mathrm{R}$	$H_0^{\mathrm{FR}}.\mathrm{R}$
CI	95%	95%	95%	95%	95%

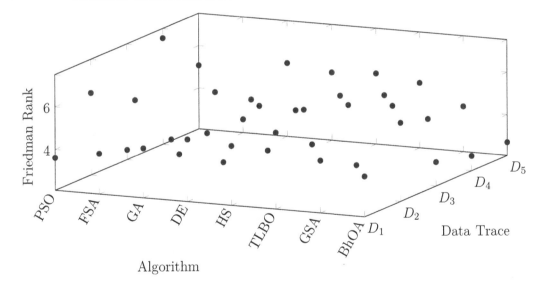

FIGURE 4.10 Friedman test ranks of metaheuristic algorithms based prediction models

4.5 STATISTICAL PERFORMANCE ASSESSMENT

Based on the experimental findings as presented in the above sections, it is very difficult to observe any conclusive remarks about the performance of the algorithms. The statistical analysis helps to find the presence of significant differences in the performance if there is any. Therefore, an in-depth analysis is conducted using the Friedman test with Finner post-hoc analysis. This test compares the pairwise performance of the algorithms. The Friedman test works around a null hypothesis (H_0^{FR}) which believes that the mean results of every candidate under test are the same. Table 4.1 and Fig. 4.10 show the analytical results which clearly depict the presence of a significant difference in the performance of the algorithms on the forecasts of all data traces except D_1 as the test successfully rejects the null hypothesis. It can be observed from mean ranks that PSO performs better on D_1 and D_3; DE produced better forecasts on D_2 while BhOA and FSA outperformed others on D_4 and D_5. Therefore, it can be stated that PSO and DE achieved better forecast accuracy on web server workloads while BhOA and FSA produced more accurate forecasts for cloud server workloads.

The pairwise comparisons conducted using Finner post-hoc method are listed in Tables 4.2 and 4.3, where acceptance and rejection of H_0^{FN} are represented by ✓ and ✗ respectively, '-' (hyphen) represents two possible cases, first, the comparison of an algorithm is not possible with itself, second, the result is already shown in the upper triangular matrix positions of the table. The test considers that the mean of the results of both algorithms is equal for each pair. The test results observed no difference among forecasts on D_1 and H_0^{FN} was accepted. However, we observed that the H_0^{FN} was accepted for $\aleph = 0.05$ but many pairs rejected H_0^{FN} for $\aleph = 0.1$. Based on these findings, the presence of a significant difference between the results can be stated with 10% risk and 90% confidence interval. In the case of D_3; BhOA, FSA, and PSO produced different results. The BhOA and FSA were able to produce different and better predictions for D_4 and D_5.

TABLE 4.2 Finner test post-hoc analysis statistics of metaheuristic algorithms based prediction models ($\aleph = 0.05$)

A_1 vs A_2	D_1 Statistic	D_1 p-value	D_2 Statistic	D_2 p-value	D_3 Statistic	D_3 p-value	D_4 Statistic	D_4 p-value	D_5 Statistic	D_5 p-value
PSO vs FSA	0.4100	0.7520	0.1863	0.8724	0.7454	0.6555	4.8076	0.0000	2.6460	0.0421
PSO vs DE	1.5653	0.5832	3.0560	0.0610	2.9069	0.0253	3.2050	0.0125	0.2236	0.8786
PSO vs GA	0.8944	0.6317	2.8324	0.0610	3.4286	0.0128	2.8324	0.0256	0.5218	0.7613
PSO vs HS	1.4162	0.6150	2.2361	0.0694	2.9069	0.0253	2.2361	0.0694	0.3727	0.8227
PSO vs TLBO	2.3106	0.4457	1.7143	0.1770	3.3541	0.0128	2.0125	0.1086	0.0745	0.9465
PSO vs GSA	1.0435	0.6312	2.6088	0.0610	3.5032	0.0128	3.0560	0.0156	0.9690	0.4862
PSO vs BhOA	0.4100	0.7520	0.1863	0.8724	0.7454	0.6555	4.8076	0.0000	2.6460	0.0421
FSA vs DE	1.1553	0.6312	2.8696	0.0610	2.1615	0.0701	1.6025	0.1830	2.8696	0.0421
FSA vs GA	0.4845	0.7426	2.6460	0.0610	2.6833	0.0269	1.9752	0.1090	3.1678	0.0421
FSA vs HS	1.0062	0.6312	2.0497	0.0996	2.1615	0.0701	2.5715	0.0350	3.0187	0.0421
FSA vs TLBO	1.9007	0.5626	1.5280	0.2370	2.6088	0.0269	2.7951	0.0256	2.7206	0.0421
FSA vs GSA	0.6336	0.7080	2.4224	0.0610	2.7578	0.0269	1.7516	0.1533	1.6771	0.2212
FSA vs BhOA	0.0000	1.0000	0.0000	1.0000	0.0000	1.0000	0.0000	1.0000	0.0000	1.0000
DE vs GA	0.6708	0.7051	0.2236	0.8674	0.5218	0.7245	0.3727	0.7635	0.2981	0.8555
DE vs HS	0.1491	0.8994	0.8199	0.5431	0.0000	1.0000	0.9690	0.4322	0.1491	0.9170
DE vs TLBO	0.7454	0.6791	1.3416	0.2930	0.4472	0.7417	1.1926	0.3382	0.1491	0.9170
DE vs GSA	0.5218	0.7426	0.4472	0.7417	0.5963	0.7122	0.1491	0.8905	1.1926	0.3906
DE vs BhOA	1.1553	0.6312	2.8696	0.0610	2.1615	0.0701	1.6025	0.1830	2.8696	0.0421
GA vs HS	0.5218	0.7426	0.5963	0.6740	0.5218	0.7245	0.5963	0.6227	0.1491	0.9170
GA vs TLBO	1.4162	0.6150	1.1180	0.3958	0.0745	0.9577	0.8199	0.5077	0.4472	0.7914
GA vs GSA	0.1491	0.8994	0.2236	0.8674	0.0745	0.9577	0.2236	0.8563	1.4907	0.2702
GA vs BhOA	0.4845	0.7426	2.6460	0.0610	2.6833	0.0269	1.9752	0.1090	3.1678	0.0421
HS vs TLBO	0.8944	0.6317	0.5218	0.7071	0.4472	0.7417	0.2236	0.8563	0.2981	0.8555
HS vs GSA	0.3727	0.7520	0.3727	0.7779	0.5963	0.7122	0.8199	0.5077	1.3416	0.3271
HS vs BhOA	1.0062	0.6312	2.0497	0.0996	2.1615	0.0701	2.5715	0.0350	3.0187	0.0421
TLBO vs GSA	1.2671	0.6150	0.8944	0.5139	0.1491	0.9170	1.0435	0.4047	1.0435	0.4599
TLBO vs BhOA	1.9007	0.5626	1.5280	0.2370	2.6088	0.0269	2.7951	0.0256	2.7206	0.0421
GSA vs BhOA	0.6336	0.7080	2.4224	0.0610	2.7578	0.0269	1.7516	0.1533	1.6771	0.2212

TABLE 4.3 Finner test post-hoc analysis results on the null hypothesis of metaheuristic algorithms based prediction model

	Data	$\aleph = 0.05$								$\aleph = 0.01$							
		PSO	FSA	DE	GA	HS	TLBO	GSA	BhOA	PSO	FSA	DE	GA	HS	TLBO	GSA	BhOA
PSO	D_1	-	✓	✓	✓	✓	✓	✓	✓	-	✓	✓	✓	✓	✓	✓	✓
	D_2	-	✓	✓	✓	✓	✓	✓	✓	-	✓	✗	✗	✗	✓	✗	✓
	D_3	-	✓	✗	✗	✗	✗	✗	✓	-	✓	✗	✗	✗	✗	✗	✓
	D_4	-	✗	✗	✗	✓	✓	✗	✗	-	✓	✗	✗	✗	✓	✗	✗
	D_5	-	✗	✓	✓	✓	✓	✓	✗	-	✓	✓	✓	✓	✓	✓	✗
FSA	D_1	-	-	✓	✓	✓	✓	✓	✓	-	-	✓	✓	✓	✓	✓	✓
	D_2	-	-	✓	✓	✓	✓	✓	✓	-	-	✗	✗	✗	✓	✗	✓
	D_3	-	-	✓	✗	✓	✗	✗	✓	-	-	✗	✗	✗	✗	✗	✓
	D_4	-	-	✓	✓	✗	✗	✓	✓	-	-	✓	✓	✗	✗	✓	✓
	D_5	-	-	✗	✗	✗	✗	✓	✓	-	-	✗	✗	✗	✗	✓	✓
DE	D_1	-	-	-	✓	✓	✓	✓	✓	-	-	-	✓	✓	✓	✓	✓
	D_2	-	-	-	✓	✓	✓	✓	✓	-	-	-	✓	✓	✓	✓	✗
	D_3	-	-	-	✓	✓	✓	✓	✓	-	-	-	✓	✓	✓	✓	✗
	D_4	-	-	-	✓	✓	✓	✓	✓	-	-	-	✓	✓	✓	✓	✓
	D_5	-	-	-	✓	✓	✓	✓	✗	-	-	-	✓	✓	✓	✓	✗
GA	D_1	-	-	-	-	✓	✓	✓	✓	-	-	-	-	✓	✓	✓	✓
	D_2	-	-	-	-	✓	✓	✓	✓	-	-	-	-	✓	✓	✓	✗
	D_3	-	-	-	-	✓	✓	✓	✗	-	-	-	-	✓	✓	✓	✗
	D_4	-	-	-	-	✓	✓	✓	✓	-	-	-	-	✓	✓	✓	✓
	D_5	-	-	-	-	✓	✓	✓	✗	-	-	-	-	✓	✓	✓	✗
HS	D_1	-	-	-	-	-	✓	✓	✓	-	-	-	-	-	✓	✓	✓
	D_2	-	-	-	-	-	✓	✓	✓	-	-	-	-	-	✓	✓	✗
	D_3	-	-	-	-	-	✓	✓	✓	-	-	-	-	-	✓	✓	✗
	D_4	-	-	-	-	-	✓	✓	✗	-	-	-	-	-	✓	✓	✗
	D_5	-	-	-	-	-	✓	✓	✗	-	-	-	-	-	✓	✓	✗
TLBO	D_1	-	-	-	-	-	-	✓	✓	-	-	-	-	-	-	✓	✓
	D_2	-	-	-	-	-	-	✓	✓	-	-	-	-	-	-	✓	✓
	D_3	-	-	-	-	-	-	✓	✗	-	-	-	-	-	-	✓	✗
	D_4	-	-	-	-	-	-	✓	✗	-	-	-	-	-	-	✓	✗
	D_5	-	-	-	-	-	-	-	✗	-	-	-	-	-	-	-	✗
GSA	D_1	-	-	-	-	-	-	-	✓	-	-	-	-	-	-	-	✓
	D_2	-	-	-	-	-	-	-	✓	-	-	-	-	-	-	-	✗
	D_3	-	-	-	-	-	-	-	✗	-	-	-	-	-	-	-	✗
	D_4	-	-	-	-	-	-	-	✓	-	-	-	-	-	-	-	✓
	D_5	-	-	-	-	-	-	-	✓	-	-	-	-	-	-	-	✓
BhOA	D_1	-	-	-	-	-	-	-	-	-	-	-	-	-	-	-	-
	D_2	-	-	-	-	-	-	-	-	-	-	-	-	-	-	-	-
	D_3	-	-	-	-	-	-	-	-	-	-	-	-	-	-	-	-
	D_4	-	-	-	-	-	-	-	-	-	-	-	-	-	-	-	-
	D_5	-	-	-	-	-	-	-	-	-	-	-	-	-	-	-	-

CHAPTER 5

Evolutionary Neural Networks

DIFFERENTIAL EVOLUTION is one of the most reliable numerical function optimization approach introduced by Price and Storn in 1995 [105]. It uses evolutionary operators including mutation, crossover, and selection to explore the solution space for an optimal solution. The approach has proved its optimization ability among evolutionary algorithms in competitions such as IEEE evolutionary computation [106]. This chapter discusses the predictive models that use the neural networks trained using differential evolution as the underlying architecture.

5.1 NEURAL NETWORK PREDICTION FRAMEWORK DESIGN

The artificial neural networks can be effectively used to model a predictive framework. The common workflow of learning the model involves various intermediate steps such as data preprocessing, data preparation, training, testing, etc. A generic flow of learning is shown in Fig. 5.1. First, it preprocesses and prepares the data as per the model requirement. Preprocessing is a combination of aggregation and scaling. The workloads may arrive at any time in the server but any forecasting model needs the workload information at fixed time intervals due to the fact that the workload history is commonly used as time-indexed data. Therefore, the predictive model first aggregates the workload data at a fixed time interval. The aggregated workload information is scaled in $[0, 1]$ using min-max scaling. Once the data is scaled in the desired range, the data is prepared to feed in the network. Let $X = \{x_1, x_2, \ldots, x_t\}$ be the aggregated and scaled actual workload arriving to the cloud servers at a fixed time interval. Given that the predictive model uses a neural network composed of n neurons in the input layer, the data is organized as shown in eq. (5.1). The workload data is arranged such that the network will have previous n consecutive workload information to predict the workload at next time step (refer to $\varkappa$ in eq. (5.1)) and the actual workload information at next time step is used as ground truth for the corresponding input pattern (refer to ς in eq. (5.1)).

$$\varkappa = \begin{bmatrix} x_1 & x_2 & \cdots & x_n \\ x_2 & x_3 & \cdots & x_{n+1} \\ \vdots & \vdots & \ddots & \vdots \\ x_k & x_{k+1} & \cdots & x_{n+k-1} \end{bmatrix}, \varsigma = \begin{bmatrix} x_{n+1} \\ x_{n+2} \\ \vdots \\ x_{n+k} \end{bmatrix} \tag{5.1}$$

DOI: 10.1201/9781003110101-5

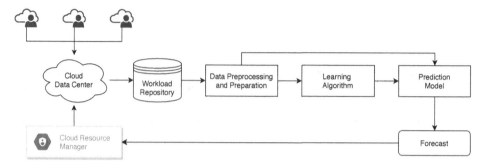

FIGURE 5.1 Neural network-based workload prediction model

Let X_R be a set of randomly generated workload values measured over a fixed interval of unit time t. Given that the network needs the workload information at unit time t only, the aggregation operation may be skipped. The next operation to be performed is the scaling of the workloads. Let X'_R be the scaled workload values and given that $n = 5$, the X'_R can be organized as $\varkappa_R$ and ς_R as shown in eq. (5.2).

$$X_R = \{39, 29, 88, 51, 5, 88, 68, 94, 85, 68, 3, 4, 11, 91, 51, 18, 70, 55, 47, 2, 7, 17, 33, 96, 78\}$$
$$X'_R = \{0.39, 0.29, 0.91, 0.52, 0.03, 0.91, 0.70, 0.98, 0.88, 0.70, 0.01, 0.02, 0.10, 0.95, 0.52,$$
$$0.17, 0.72, 0.56, 0.48, 0.00, 0.05, 0.16, 0.33, 1.00, 0.81, \}$$

$$\varkappa_R = \begin{bmatrix} 0.39 & 0.29 & 0.91 & 0.52 & 0.03 \\ 0.29 & 0.91 & 0.52 & 0.03 & 0.91 \\ 0.91 & 0.52 & 0.03 & 0.91 & 0.70 \\ 0.52 & 0.03 & 0.91 & 0.70 & 0.98 \\ 0.03 & 0.91 & 0.70 & 0.98 & 0.88 \\ \vdots & \vdots & \vdots & \vdots & \vdots \end{bmatrix}, \varsigma_R = \begin{bmatrix} 0.91 \\ 0.70 \\ 0.98 \\ 0.88 \\ 0.70 \\ \vdots \end{bmatrix} \tag{5.2}$$

The predictive models discussed in this chapter use the feed-forward neural network as an underlying architecture which is composed of $n - p - q$ neurons in such a way that it has n input neurons, p hidden neurons, and q output neurons. Since the network has to predict a real-valued output, the output layer will have only one neuron. The network is composed of ten input neurons and seven hidden neurons. The learning periods for mutation and crossover operators are ten iterations. The maximum iterations to execute the learning algorithm is 250 and 60% of training data is used to learn the network weights. Another important design parameter for a neural network-based model is the choice of activation function across the network. In this chapter, the models are using a combination of linear and sigmoid activation functions at different layers. The rule for the choice of the activation function is given in eq. (5.3), where z is a weighted sum of the inputs a neuron receives from a number of neurons of its previous layer.

$$f(z) = \begin{cases} z & \text{if input layer} \\ \frac{1}{1+e^{-z}} & \text{otherwise} \end{cases} \quad (5.3)$$

The predictive models use a ratio of 60:40 between the training and testing data. We know that the training data is given to the network with corresponding ground truth i.e. actual workload values at the next time step and the network learns the synaptic connection weights using that data. Once the learning of the network is over, it uses the test data to assess its learning performance. The difference between training and testing data is that test data is given to the network without ground truth whereas the training data carries the corresponding desired outcome with itself. Moreover, the performance of the networks used in this chapter is assessed using root mean squared error.

5.2 NETWORK LEARNING

Artificial neural networks have revolutionized the way of learning across various applications including pattern recognition, speech recognition, computer vision, data analysis, and many others [2, 22, 42, 131]. The usage of neural networks allows us to mimic the learning behavior of human beings. In the modern world of full of smart applications, neural networks are looked at as an expert machines (at least in the domain of their expertise) but each expert needs to learn at least once. The neural networks are not exceptions to this and they get the training to become experts at any given task.

It is a very difficult and challenging task to learn the mapping between history and future. The gradient-based solutions are not always the first choice to train the network if the problem is very complex such as NP-Hard. Given that the search space is multi-modal, the population-based search techniques are one of the possible solutions to address the issues of gradient-based solutions. Among population-based search methods, differential evolution is a very simple and reliable search technique. Similar to any population-based search techniques the differential evolution also works around a set of solutions. Here, a solution is composed of a set of real-valued numbers that represent the network synaptic connection weights. A set of N solutions are generated randomly using $s_{i,j} = lb_j + r \times (ub_j - lb_j)$. The length of the solutions can be computed as $p(n + 2)$ (including bias weights).

The optimizer explores the search space by generating new solutions by the means of recombination and selection from the initially generated random solutions. First, it selects three distinct solutions and adds one to the weighted difference of the other two solutions. This process is referred to as mutation and it ensures diversity and prevents premature convergence. Moreover, it also prevents the network to stuck in local optima. The mutant solution (generated from mutation) performs the crossover with the base solution and offsprings are generated. These offspring solutions are evaluated on the same objective function and their fitness is compared with the existing pool of solutions. If any fitter solution is observed in the pool of offspring, it replaces the less fit solution in the population i.e. the rule of survival of the fittest is followed. This chapter explores the predictive models which are trained using two modified versions of the differential evolution. The modified algorithms extend the self-adaptive differential evolution [107] to develop more robust algorithms.

5.3 RECOMBINATION OPERATOR STRATEGY LEARNING

The recombination operators play a critical role in the process of searching for an optimal solution in a very complex search space. The quality of approximated solution highly depends on the solutions generated using recombination operators. A wide number of operations have been proposed to generate offsprings. Moreover, it has been observed that these operators perform differently across the applications. Thus, it is very important to select the best suitable operator to achieve a good approximation.

5.3.1 Mutation Operator

The differential evolution first applies the mutation operation to generate offspring solutions. The mutation operator ensures the diversity, prevents premature convergence, and also avoids the local optima. A number of different operators are available for the mutation. The first discussed approach learns the suitable mutation operator among *DE/current to best/1*, *DE/best/1*, and *DE/rand/1* [74].

5.3.1.1 DE/current to best/1

Let s_i be the base or current solution, the *DE/current to best/1* mutation selects two distinct solutions randomly such that $i \neq r_1 \neq r_2$. The weighted differences between best (s_{best}) and current solutions (s_i), and between both randomly selected solutions $(s_{r_1}$ and $s_{r_2})$ are added to the current solution as shown in eq. (5.4). This operator focuses on faster convergence.

$$v_i = s_i + F \times (s_{best} - s_i) + F \times (s_{r_1} - s_{r_2}) \tag{5.4}$$

5.3.1.2 DE/best/1

The *DE/best/1* is a good choice to use in solving optimization problems [93]. It computes the weighted difference of two distinct and randomly selected solutions and adds it to the best solution achieved so far as shown in eq. (5.5).

$$v_i = s_{best} + F \times (s_{r_1} - s_{r_2}) \tag{5.5}$$

5.3.1.3 DE/rand/1

It is one of the most used mutation operators across applications. It selects three distinct random solutions such that $r_1 \neq r_2 \neq r_3$ from the pool of the current population. The weighted difference between two random solutions is added to the third solution as shown in eq. (5.6). It has been observed that the operator *DE/rand/1* is good at diversity maintenance of the population [107].

$$v_i = s_{r_3} + F_i \times (s_{r_1} - s_{r_2}) \tag{5.6}$$

5.3.2 Crossover Operator

Crossover is another recombination operator that allows generating offspring using two randomly selected solutions from the pool of existing solutions. The crossover operator is adapted from theory of evolution where two individual participates to generate new individuals. In differential evolution, the parents are selected from the pool of population members and mutants. Again the choice of the crossover operator highly impacts the quality of solutions. Biphase adaptive differential evolution (BaDE) also learns the best suitable crossover operator along with the mutation operator [72].

5.3.2.1 Ring Crossover

In this operator, both parents are connected and organized in such a way that they form a ring structure as shown in Fig. 5.2. A random number in [1, D] is generated to mark a cut point. This cut point is treated as an origin point to generate both offsprings. The first offspring is generated by selecting D values clockwise from the origin point and the second offspring selects D values anticlockwise. The ring crossover uses the swap and reverses effect which helps in maintaining the diversity of the population [56]. Also, the ring crossover operator better explores the search space and avoids premature convergence [67].

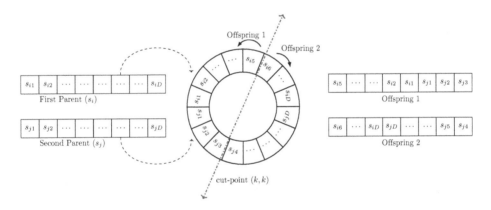

FIGURE 5.2 Ring crossover

Let $s_i = \{s_{i1}, s_{i2}, s_{i3}, \ldots, s_{iD}\}$ and $s_j = \{s_{j1}, s_{j2}, s_{j3}, \ldots, s_{jD}\}$ be two individuals selected to participate in the crossover. Both individuals are connected by their endpoints and form the ring. A random number $\Re \in [1, D]$ is generated to mark the cut point and two offsprings u_i and u_j are generated.

5.3.2.2 Heuristic Crossover

Heuristic crossover operator takes the fitness values into consideration for guiding the search towards the desired region. This operator generates the offspring such that the newly generated solution is closer to the individual with better fitness which helps in improving the average health of the offspring [132]. Moreover, it is observed that the heuristic crossover helps to choose better weights of the network than various other operators such as arithmetic crossover [3]. The graphical representation of the heuristic crossover is shown in Fig. 5.3, where star (*) indicates the parent with better fitness.

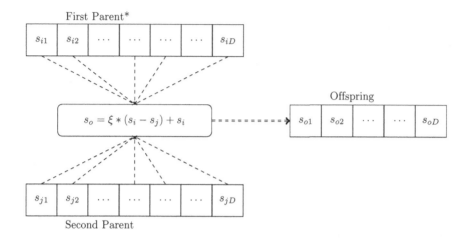

FIGURE 5.3 Heuristic crossover

Let s_i and v_i be two parents such that $s_i = \{s_{i1}, s_{i2}, s_{i3}, \ldots, s_{iD}\}$ and $v_i = \{v_{i1}, v_{i2}, v_{i3}, \ldots, v_{iD}\}$. Assuming that s_i has better fitness (s_i*) value than v_i i.e. $f_{s_i} \leq f_{v_i}$ (for minimization problem). Then the offspring solution (u_i) is generated using eq. (5.7), where ξ is a randomly generated number in the range $[0, 1]$.

$$u_i = \xi \times (s_i - v_i) + s_i \tag{5.7}$$

5.3.2.3 Uniform Crossover

The next crossover operator is the uniform crossover operator which exchanges the genetic information at the gene level as opposed to other common crossover operators that work on segment level [101]. In this operation, each gene exchanges the values with the probability of CR as shown in Fig. 5.4.

5.3.3 Operator Learning Process

The adaptation process in two different variants of modified differential evolution is applied. First variant, self-adaptive differential evolution (SaDE) which selects the best suitable mutation operator whereas the second variant, biphase adaptive differential evolution (BaDE) selects the best suitable mutation and crossover also. The experiments are conducted using *DE/current to best/1*, *DE/rand/1*, and *DE/best/1* mutation operators on the basis of their functionalities [93, 107]. Similarly, *heuristic crossover*, *ring crossover*, and *uniform crossover* are selected for the experimentation.

Let P_1^Z, P_2^Z, and P_3^Z be the probabilities to apply *DE/rand/1*, *DE/current to best/1*, and *DE/best/1* respectively. Similarly, P_1^Q, P_2^Q, and P_3^Q are the probabilities to apply ring crossover, heuristic crossover, and uniform crossover respectively. In

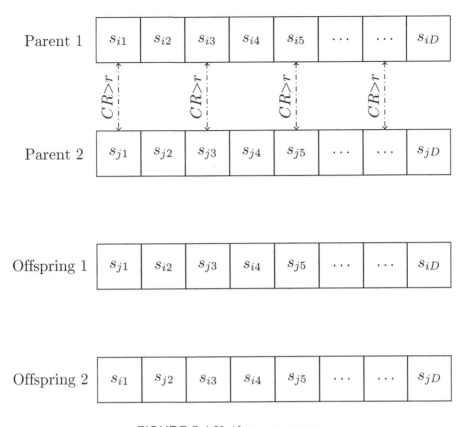

FIGURE 5.4 Uniform crossover

beginning, each operator is selected with equal chances i.e. $P_1^{\mathcal{Z}} = 0.33$, $P_2^{\mathcal{Z}} = 0.33$, and $P_3^{\mathcal{Z}} = 0.34$. Similarly, the probabilities for the crossover operators are selected as $P_1^{\mathcal{Q}} = 0.33$, $P_2^{\mathcal{Q}} = 0.33$, and $P_3^{\mathcal{Q}} = 0.34$ to provide equal chances to each of the operators. Every individual carries two selection probabilities to select mutation and crossover strategy. In the beginning, the selection probabilities are randomly generated in (0, 1) and stored in two vectors named $\mathcal{Z}_P$ and $\mathcal{Q}_P$ which store the selection probabilities associated with mutation and crossover operators. Since each solution has one value for mutation and crossover, the length of $\mathcal{Z}_P$ and $\mathcal{Q}_P$ is N. For a base solution s_i, the mutation and crossover operators are selected as shown in eqs. (5.8) and (5.9) respectively, where MS_i and CS_i are the mutation and crossover operators which will be applied to generate u_i.

$$MS_i = \begin{cases} DE/rand/1 & 0 < \mathcal{Z}_P(i) \le P_1^{\mathcal{Z}} \\ DE/\text{current to best}/1 & P_1^{\mathcal{Z}} < \mathcal{Z}_P(i) \le (P_1^{\mathcal{Z}} + P_2^{\mathcal{Z}}) \\ DE/best/1 & otherwise \end{cases} \qquad (5.8)$$

$$CS_i = \begin{cases} \text{Ring Crossover} & 0 < \mathcal{Q}_P(i) \leq P_1^{\mathcal{Q}} \\ \text{Heuristic Crossover} & P_1^{\mathcal{Q}} < \mathcal{Q}_P(i) \leq (P_1^{\mathcal{Q}} + P_2^{\mathcal{Q}}) \\ \text{Uniform Crossover} & otherwise \end{cases} \qquad (5.9)$$

At every iteration, the optimization process keeps the records of the offspring solutions both entering into the next iteration and failing in reaching to next iteration. The count of successful offspring candidates generated by each of the mutation operators is recorded in $S_1^{\mathcal{Z}}$, $S_2^{\mathcal{Z}}$, and $S_3^{\mathcal{Z}}$ respectively. Similarly, the count of failed offspring solutions generated by each of the crossover operators is recorded in $F_1^{\mathcal{Z}}$, $F_2^{\mathcal{Z}}$, and $F_3^{\mathcal{Z}}$ respectively. The number of successful and failed offspring solutions generated by each crossover operator is stored in $S_1^{\mathcal{Q}}$, $S_2^{\mathcal{Q}}$, $S_3^{\mathcal{Q}}$, $F_1^{\mathcal{Q}}$, $F_2^{\mathcal{Q}}$, and $F_3^{\mathcal{Q}}$ respectively. After a certain number of iteration i.e. learning period, the probabilities of mutation and crossover operators are updated using eqs. (5.10) to (5.15). After updating the operator selection probabilities associated with each variant, the count of successful and unsuccessful offsprings in reaching into next iteration is reset i.e. the values of $S_i^{\mathcal{Z}}$, $F_i^{\mathcal{Z}}$, $S_i^{\mathcal{Q}}$, and $F_i^{\mathcal{Q}}$ are assigned to zero. The effect of the different values of both learning periods is modeled and results are shown in Fig. 5.5 which shows that different learning rates suit different prediction interval-based forecasting.

$$P_1^{\mathcal{Z}} = \frac{S_1^{\mathcal{Z}}(S_2^{\mathcal{Z}} + F_2^{\mathcal{Z}} + S_3^{\mathcal{Z}} + F_3^{\mathcal{Z}})}{2(S_2^{\mathcal{Z}}S_3^{\mathcal{Z}} + S_1^{\mathcal{Z}}S_3^{\mathcal{Z}} + S_1^{\mathcal{Z}}S_2^{\mathcal{Z}}) + F_1^{\mathcal{Z}}(S_2^{\mathcal{Z}} + S_3^{\mathcal{Z}}) + F_2^{\mathcal{Z}}(S_1^{\mathcal{Z}} + S_3^{\mathcal{Z}}) + F_3^{\mathcal{Z}}(S_1^{\mathcal{Z}} + S_2^{\mathcal{Z}})} \qquad (5.10)$$

$$P_2^{\mathcal{Z}} = \frac{S_2^{\mathcal{Z}}(S_1^{\mathcal{Z}} + F_1^{\mathcal{Z}} + S_3^{\mathcal{Z}} + F_3^{\mathcal{Z}})}{2(S_2^{\mathcal{Z}}S_3^{\mathcal{Z}} + S_1^{\mathcal{Z}}S_3^{\mathcal{Z}} + S_1^{\mathcal{Z}}S_2^{\mathcal{Z}}) + F_1^{\mathcal{Z}}(S_2^{\mathcal{Z}} + S_3^{\mathcal{Z}}) + F_2^{\mathcal{Z}}(S_1^{\mathcal{Z}} + S_3^{\mathcal{Z}}) + F_3^{\mathcal{Z}}(S_1^{\mathcal{Z}} + S_2^{\mathcal{Z}})} \qquad (5.11)$$

$$P_3^{\mathcal{Z}} = 1 - (P_1^{\mathcal{Z}} + P_2^{\mathcal{Z}}) \qquad (5.12)$$

$$P_1^{\mathcal{Q}} = \frac{S_1^{\mathcal{Q}}(S_2^{\mathcal{Q}} + F_2^{\mathcal{Q}} + S_3^{\mathcal{Q}} + F_3^{\mathcal{Q}})}{2(S_2^{\mathcal{Q}}S_3^{\mathcal{Q}} + S_1^{\mathcal{Q}}S_3^{\mathcal{Q}} + S_1^{\mathcal{Q}}S_2^{\mathcal{Q}}) + F_1^{\mathcal{Q}}(S_2^{\mathcal{Q}} + S_3^{\mathcal{Q}}) + F_2^{\mathcal{Q}}(S_1^{\mathcal{Q}} + S_3^{\mathcal{Q}}) + F_3^{\mathcal{Q}}(S_1^{\mathcal{Q}} + S_2^{\mathcal{Q}})} \qquad (5.13)$$

$$P_2^{\mathcal{Q}} = \frac{S_2^{\mathcal{Q}}(S_1^{\mathcal{Q}} + F_1^{\mathcal{Q}} + S_3^{\mathcal{Q}} + F_3^{\mathcal{Q}})}{2(S_2^{\mathcal{Q}}S_3^{\mathcal{Q}} + S_1^{\mathcal{Q}}S_3^{\mathcal{Q}} + S_1^{\mathcal{Q}}S_2^{\mathcal{Q}}) + F_1^{\mathcal{Q}}(S_2^{\mathcal{Q}} + S_3^{\mathcal{Q}}) + F_2^{\mathcal{Q}}(S_1^{\mathcal{Q}} + S_3^{\mathcal{Q}}) + F_3^{\mathcal{Q}}(S_1^{\mathcal{Q}} + S_2^{\mathcal{Q}})} \qquad (5.14)$$

$$P_3^{\mathcal{Q}} = 1 - (P_1^{\mathcal{Q}} + P_2^{\mathcal{Q}}) \qquad (5.15)$$

In differential evolution, there are another two critical parameters i.e. crossover and mutation rates. The separate crossover rate is initialized for every individual.

The initial values of crossover rates are normally distributed such that the range of values is $(0, 1]$, the mean value is (CR_μ) 0.5, and the standard deviation (CR_σ) is 0.1. Similarly, the mutation rate (F) values are initialized with the standard deviation (F_σ) is 0.3.

5.4 ALGORITHMS AND ANALYSIS

Both variants of the differential evolution optimization algorithm optimally learn the best suitable version of the recombination operators. The first variant i.e. SaDE (Algorithm 5.1) finds the best suitable mutation operator. Whereas the second variant of the algorithm i.e. BaDE (5.2) applies the adaption at two levels meaning that the algorithm finds the best suitable mutation and crossover operators. Let $\mathcal{O}(1)$ be the complexity of one random number generation, the $\mathcal{O}(ND)$ is the time complexity of population initialization as it involves generating N vectors, each of length D. Since the value of D is obtained by $p(n + 2)$, the effective time complexity of population initialization becomes $\mathcal{O}(Npn)$. Furthermore, the $p < n$ (at least for a high dimensional data) and thus the time complexity may be represented as $\mathcal{O}(n^2 N)$. The next step of the algorithm is to generate three vectors $(\mathcal{Q}_v, \mathcal{Z}_v, \mathcal{Z}_P)$ of random numbers, each of size N and it adds $\mathcal{O}(N)$ complexity. In any population-based search optimization algorithm, the fitness assessment is one of the great sources of complexity. Similarly, the fitness assessment in SaDE accounts for a large amount of complexity. Let forecasting a single data point which takes $\mathcal{O}(n^2)$ be the unit operation in the fitness assessment. Since one network is assessed on the forecasts of m data points and the population consists of N networks, the complexity of fitness assessment operation becomes $\mathcal{O}(n^2 mN)$. The generation of an offspring solution using recombination operators consumes $\mathcal{O}(n^2)$, thus, the generation of N offspring solutions needs $\mathcal{O}(n^2 N)$. Another key operation in the algorithm is the selection which consumes $\mathcal{O}(N)$. The total complexity of the algorithm becomes $\mathcal{O}(Gmn^2 N)$ as the recombination operators, offspring fitness assessment, and selection are repeated for G times. The detailed analysis can be found in [74].

The second variant of the algorithm (Algorithm 5.2) uses some additional instructions to incorporate the additional feature of learning the best suitable crossover operator. The additional computations are executed for a vector $(\mathcal{Q}_P)$ generation of N random numbers and some additional checkpoints in learning the best suitable operator. However, the additional computations do not affect the total complexity of the algorithm. The readers should refer to [72] for a detailed discussion on the complexity and algorithm.

Algorithm 5.1 Self-adaptive differential evolution based forecasting framework pseudocode [74]

1: Initialize $CR_\mu = F_\mu = 0.5$, $CR_\sigma = 0.1$, $F_\sigma = 0.3$, $P_1^{\mathcal{Z}} = P_2^{\mathcal{Z}} = 0.33$, $P_3^{\mathcal{Z}} = 0.34$
2: Randomly initialize N networks of length D (population) /* Here D is number of connections in network */
3: Generate crossover CR_v and mutation F_v vectors (length N) CR_v, $F_v \in (0,1]$
4: Evaluate each network on training data using objective function
5: **repeat**
6:   **for** each generation g **do**
7:     Generate a vector $\mathcal{Z}_P$ of N random numbers $\in (0,1]$
8:     **for** each solution i **do**
9:       Generate $r_1 \neq r_2 \neq r_3 \neq i \in [1, N]$ and $j_{rand} \in [1, D]$
10:       **if** $0 < \mathcal{Z}_P(i) \leq P_1^{\mathcal{Z}}$ **then**
11:         Apply $DE/rand/1$ mutation strategy and crossover
12:       **else**
13:         **if** $P_1^{\mathcal{Z}} < \mathcal{Z}_P(i) \leq (P_1^{\mathcal{Z}} + P_2^{\mathcal{Z}})$ **then**
14:           Apply $DE/current\ to\ best/1$ mutation strategy and crossover
15:         **else**
16:           Apply $DE/best/1$ mutation strategy and crossover
17:         **end if**
18:       **end if**
19:     **end for**
20:     Evaluate offspring vectors using objective function
21:     Select participants for next generation from offspring vectors and population
22:     Update values of $S_1^{\mathcal{Z}}, S_2^{\mathcal{Z}}, S_3^{\mathcal{Z}}, F_1^{\mathcal{Z}}, F_2^{\mathcal{Z}}$, and $F_3^{\mathcal{Z}}$
23:     Update $P_1^{\mathcal{Z}}, P_2^{\mathcal{Z}}$, and $P_3^{\mathcal{Z}}$ /* (After every lp^F generations) */
24:     Regenerate CR_v /* (After every lp^{CR} generations) */
25:     Recalculate CR_μ /* (After every g_{lp}^{CR} generations) */
26:   **end for**
27: **until** termination criteria is not met

Algorithm 5.2 Biphase adaptive differential evolution based forecasting framework pseudocode [72]

1: Initialize $CR_\mu = F_\mu = 0.5$, $CR_\sigma = 0.1$, $F_\sigma = 0.3$, $P_1^\mathcal{Z} = P_2^\mathcal{Z} = 0.33$, $P_3^\mathcal{Z} = 0.34$, $P_1^\mathcal{Q} = P_2^\mathcal{Q} = 0.33$, $P_3^\mathcal{Q} = 0.34$

2: Randomly initialize N networks of length D (population) /* Here D is number of connections in network */

3: Generate crossover CR_v and mutation F_v vectors (length N) CR_v, $F_v \in (0,1]$

4: Evaluate each network on training data using objective function

5: **repeat**

6:   **for** each generation g **do**

7:     Generate vectors $\mathcal{Z}_P$ and $\mathcal{Q}_P$ of N random numbers $\in (0,1]$

8:     **for** each solution i **do**

9:       Generate $r_1 \neq r_2 \neq r_3 \neq i \in [1, N]$

10:       **if** $0 < \mathcal{Z}_P(i) \leq P_1^\mathcal{Z}$ **then**

11:         $v_i^g = s_{r_3}^g + F_i \times (s_{r_1}^g - s_{r_2}^g)$

12:       **else**

13:         **if** $P_1^\mathcal{Z} < \mathcal{Q}_P(i) \leq (P_1^\mathcal{Z} + P_2^\mathcal{Z})$ **then**

14:           $v_i^g = s_i^g + F_i \times (s_{best}^g - s_i^g) + F_i \times (s_{r_1}^g - s_{r_2}^g)$

15:         **else**

16:           $v_i^g = s_{best}^g + F_i \times (s_{r_1}^g - s_{r_2}^g)$

17:         **end if**

18:       **end if**

19:     **end for**

20:     **for** each solution i **do**

21:       Generate $j_{rand} \in [1, D]$

22:       **if** $0 < \mathcal{Q}_P(i) \leq P_1^\mathcal{Q}$ **then**

23:         Apply ring crossover

24:       **else**

25:         **if** $P_1^\mathcal{Q} < \mathcal{Q}_P(i) \leq (P_1^\mathcal{Q} + P_2^\mathcal{Q})$ **then**

26:           Apply heuristic crossover

27:         **else**

28:           Apply uniform crossover

29:         **end if**

30:       **end if**

31:     **end for**

32:     Evaluate offspring vectors using objective function

33:     Select participants for next generation from offspring vectors and population

34:     Update values of $S_1^\mathcal{Z}, S_2^\mathcal{Z}, S_3^\mathcal{Z}, F_1^\mathcal{Z}, F_2^\mathcal{Z}, F_3^\mathcal{Z}, S_1^\mathcal{Q}, S_2^\mathcal{Q}, S_3^\mathcal{Q}, F_1^\mathcal{Q}, F_2^\mathcal{Q}$, and $F_3^\mathcal{Q}$

35:     Update $P_1^\mathcal{Z}, P_2^\mathcal{Z}$ and $P_3^\mathcal{Z}$ /* (After every lp^F generations) */

36:     Update $P_1^\mathcal{Q}, P_2^\mathcal{Q}$ and $P_3^\mathcal{Q}$ /* (After every lp^{CR} generations) */

37:     Regenerate CR_v /* (After every lp^{CR} generations) */

38:     Recalculate CR_μ /* (After every glp^{CR} generations) */

39:   **end for**

40: **until** termination criteria is not met

5.5 FORECAST ASSESSMENT

The performance of the predictive frameworks that use SaDE and BaDE learning algorithms to optimize the network weights are assessed on a variety of experiments. First, the network structure i.e. the number of neurons in the input and hidden layers are selected. In order to select the number of input neurons, the networks with 5 to 20 input neurons are created and their performance is assessed on a subset of data-trace. The network with 10 input neurons performed better than other networks. Thus, the network will have 10 input neurons. Furthermore, the network will have the hidden neurons equal to the $\lceil 2/3 \rceil$ input neurons. The two other parameters i.e. the population size and the maximum number of iterations or generation are fixed prior to the experiments. The experiments are conducted with a 20 member population and each experiment was executed for 250 iterations. The learning periods for updating the mutation and crossover probabilities are 10 iterations meaning that every 10 iterations the respective probabilities of the recombination operators will be updated [74].

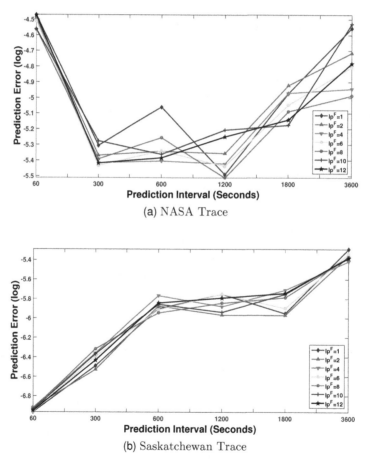

(a) NASA Trace

(b) Saskatchewan Trace

FIGURE 5.5 Learning period effect on forecast accuracy of self-adaptive differential evolution algorithm based workload prediction model [74]

5.5.1 Short Term Forecast

The performance of both predictive frameworks is accessed on data traces of NASA Trace and Saskatchewan Trace. First, the forecast results obtained on a 1-minute prediction interval by SaDE based predictive framework are shown in Fig. 5.6 which includes the results depicting the actual and predictive workload values on an entire and a small subset of both data traces. One might easily notice the closeness of actual and predicted workload values by observing the visual results. The predictive framework is able to model the patterns of actual workloads and forecasts the workloads reasonably closer to the actual workload. Furthermore, the forecast residuals are depicted in Fig. 5.7 corresponding to the predicted workloads shown in Figs. 5.6c and 5.6d respectively. The residuals are very low and random in nature which means the predictive framework was able to model the workload patterns significantly.

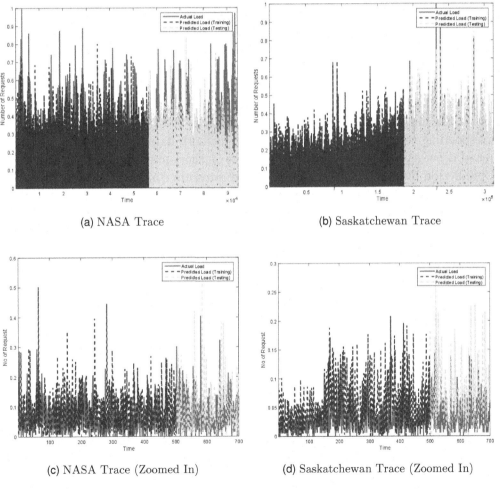

(a) NASA Trace

(b) Saskatchewan Trace

(c) NASA Trace (Zoomed In)

(d) Saskatchewan Trace (Zoomed In)

FIGURE 5.6 Short term forecast assessment of self-adaptive differential evolution algorithm based prediction model on 1-minute prediction interval [74]

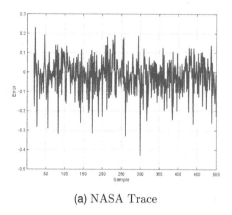

(a) NASA Trace

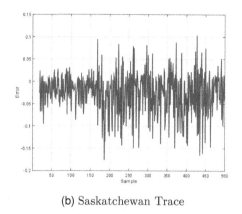

(b) Saskatchewan Trace

FIGURE 5.7 Short term forecast residuals of self-adaptive differential evolution algorithm based prediction model on 1-minute prediction interval [74]

The same experiments are conducted with the predictive framework based on BaDE algorithm. The forecast results of both data traces for a prediction interval of 1 minute are shown in Fig. 5.8. Here also, the results include the actual and predicted workloads on entire and subsets of both data traces. The close observations reveal that the forecast accuracy of the later framework is better, thus, the crossover operator learning has improved the learning ability of the algorithm significantly. The corresponding residuals are shown in Fig. 5.9 for both data traces. Furthermore, the forecast accuracy of both predictive frameworks on short-term forecasts i.e. the forecast on prediction intervals of 1, 5, 10, and 20 minutes are compared (Fig. 5.10), where the forecast accuracy is measured using RMSE. Thus, the lower values are preferred over higher values and BaDE learning algorithm-based solution produced lower errors in forecasts except for the prediction interval of 1 minute. The performance of both approaches is also compared on other parameters such as the number of iterations elapsed in the convergence (no change in the solution across successive iterations) and the time elapsed in training as shown in Table 5.1. The later approach converged faster but it took relatively larger time due to the fact that it involves additional computations in optimizing the crossover operator. Thus, it is found that the second approach performed better than the first approach on short-term forecasts.

5.5.2 Long Term Forecast

Apart from the short-term forecasts, the forecast accuracy of both approaches is assessed on long-term forecasts also. Here, if the forecast interval is 30 minutes or more then it is a long-term forecast. For the experimentation purpose, the forecasts are assessed on 30 and 60 minutes forecasts. The actual and predicted workloads obtained from SaDE based framework are shown in Fig. 5.11 including the entire and subsets of data traces. As opposed to the short-term forecasts, the approach detects and models the workload patterns more accurately as shown in Fig. 5.12. Similarly, the forecast results obtained using BaDE based model are shown in Fig. 5.13 and

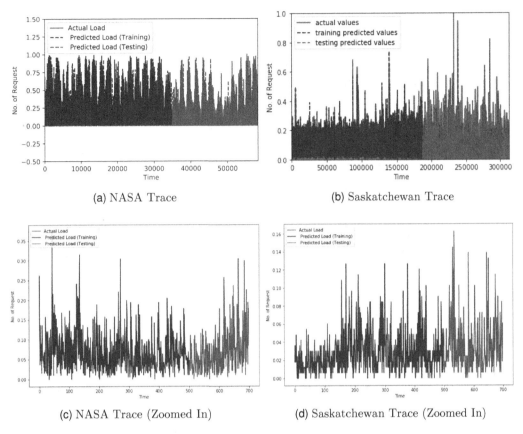

(a) NASA Trace

(b) Saskatchewan Trace

(c) NASA Trace (Zoomed In)

(d) Saskatchewan Trace (Zoomed In)

FIGURE 5.8 Short term forecast assessment of biphase adaptive differential evolution algorithm based prediction model on 1-minute prediction interval [74]

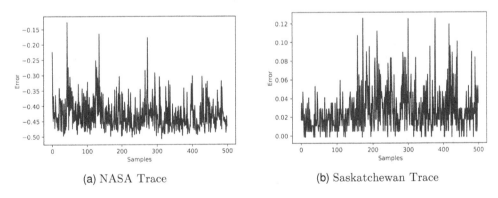

(a) NASA Trace

(b) Saskatchewan Trace

FIGURE 5.9 Short term forecast residual of biphase adaptive differential evolution algorithm based prediction model on 1-minute prediction interval [74]

corresponding residuals can be seen in Fig. 5.14. Furthermore, the performance of both models is compared on long-term forecasts and the comparison can be seen in Fig. 5.15. The BaDE model generates the forecasts with better accuracy (lower

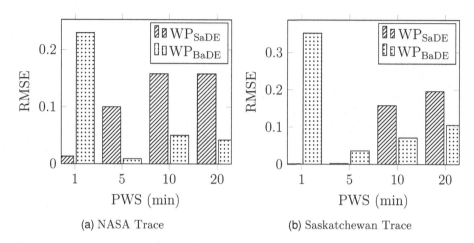

(a) NASA Trace   (b) Saskatchewan Trace

FIGURE 5.10 Comparing short term forecast errors of SaDE and BaDE based predictive frameworks

TABLE 5.1 Number of iterations and time elapsed in the training of differential evolution based predictive models for short term forecasts

PWS (min)	#Iterations				Training Time (sec)			
	NASA Trace		Saskatchewan Trace		NASA Trace		Saskatchewan Trace	
	SaDE	BaDE	SaDE	BaDE	SaDE	BaDE	SaDE	BaDE
1	47	21	61	27	121.368	771.250	661.690	14506.710
5	39	23	77	23	24.112	237.210	132.666	1566.094
10	51	23	51	23	12.144	145.464	64.807	718.919
20	47	24	51	22	6.354	69.989	33.376	339.443

values of RMSE are preferred over higher values). Similar to the short-term forecast observations, both models show a similar pattern in the number of iterations elapsed in the convergence and the time required for training (see Table 5.2). Thus, based on the observations, the BaDE learning algorithm-based framework reduces the forecast errors significantly over the SaDE learning algorithm-based framework.

TABLE 5.2 Number of iterations and time elapsed in the training of differential evolution based predictive models for long term forecasts

PWS (min)	#Iterations				Training Time (sec)			
	NASA Trace		Saskatchewan Trace		NASA Trace		Saskatchewan Trace	
	SaDE	BaDE	SaDE	BaDE	SaDE	BaDE	SaDE	BaDE
30	26	24	42	23	4.274	46.285	21.429	245.060
60	26	23	21	22	2.908	23.183	10.733	236.655

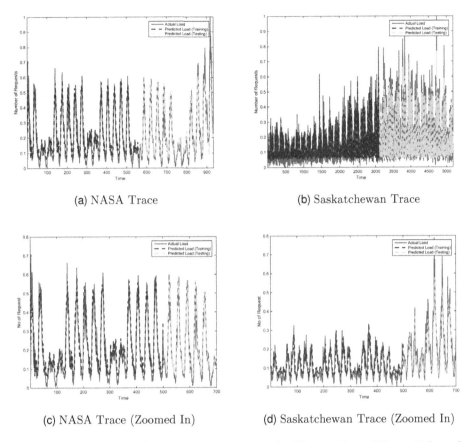

(a) NASA Trace

(b) Saskatchewan Trace

(c) NASA Trace (Zoomed In)

(d) Saskatchewan Trace (Zoomed In)

FIGURE 5.11 Long term forecast assessment of self-adaptive differential evolution algorithm based prediction model on 60-minute prediction interval

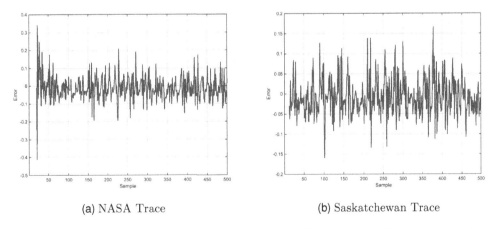

(a) NASA Trace

(b) Saskatchewan Trace

FIGURE 5.12 Long term forecast residuals of self-adaptive differential evolution algorithm based prediction model on 60-minute prediction interval

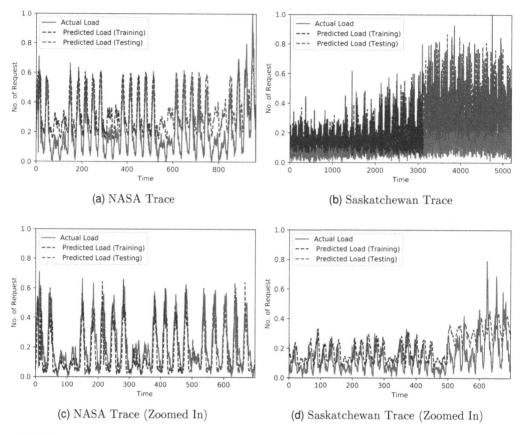

(a) NASA Trace

(b) Saskatchewan Trace

(c) NASA Trace (Zoomed In)

(d) Saskatchewan Trace (Zoomed In)

FIGURE 5.13 Long term forecast assessment of biphase adaptive differential evolution algorithm based prediction model on 60-minute prediction interval

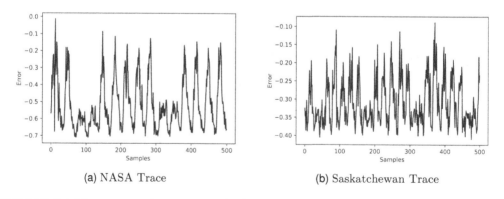

(a) NASA Trace

(b) Saskatchewan Trace

FIGURE 5.14 Long term forecast residuals of biphase adaptive differential evolution algorithm based prediction model on 60-minute prediction interval

5.6 COMPARATIVE ANALYSIS

This section compares the performance of both approaches with state-of-art methods predictive frameworks. Three different parameters are used to compare the performance

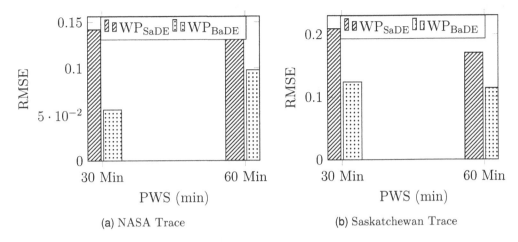

(a) NASA Trace   (b) Saskatchewan Trace

FIGURE 5.15 Comparing long term forecast errors of SaDE and BaDE based predictive frameworks

which are forecast accuracy, the time elapsed in training, and the number of iterations elapsed in reaching convergence. The state-of-art methods are based on the average, maximum, and backpropagation algorithms [104]. The first two approaches are very simple as they forecast the workload the same as the average and maximum workload received so far. Whereas the third approach uses a neural network that learns the network weights using a backpropagation algorithm.

Figure 5.16 compares the forecast accuracy (measured using RMSE) on NASA Trace. The SaDE and BaDE based predictive frameworks are better than other models. For instance, the SaDE based model successfully reduced the RMSE up to 96.06%, 98.38%, and 94.92% (relative) over maximum, average, and backpropagation-based models. Similarly, the BaDE based model relatively reduced forecast errors up to 97.19%, 98.42%, and 97.11% over maximum, average, and backpropagation-based models. Moreover, the BaDE based predictive framework improves the forecast accuracy by reducing the RMSE up to 91.00% over SaDE based model.

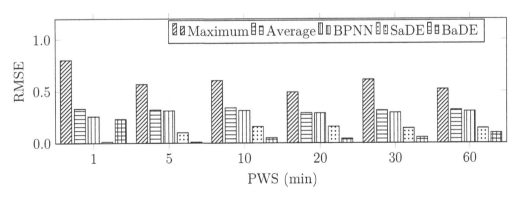

FIGURE 5.16 Comparing forecast accuracy of Maximum, Average, BPNN, SaDE, BaDE based predictive frameworks on NASA Trace

A similar comparison is performed on the forecast accuracy of Saskatchewan Trace. The relative improvements achieved by SaDE and BaDE based models are 99.53%, 99.85%, 99.40%, 90.06%, 95.34%, and 89.29% over maximum, average, and backpropagation methods as shown in Fig. 5.17. Furthermore, the BaDE based model improves the forecast accuracy over SaDE based model by reducing the RMSE up to 55.06%. Thus, adaptive learning has helped in improving forecast accuracy.

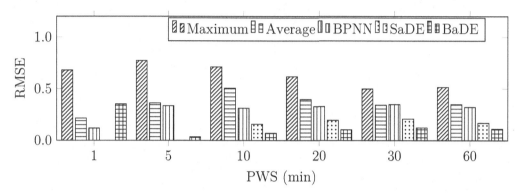

FIGURE 5.17 Comparing forecast accuracy of Maximum, Average, BPNN, SaDE, BaDE based predictive frameworks on Saskatchewan Trace

TABLE 5.3 Iterations elapsed for the training of predictive models using backpropagation, self-adaptive differential evolution, and biphase adaptive differential evolution

PWS (min)	NASA Trace			Saskatchewan Trace		
	WP_{BPNN}	WP_{SaDE}	WP_{BaDE}	WP_{BPNN}	WP_{SaDE}	WP_{BaDE}
1	250	47	21	250	61	27
5	250	39	23	250	77	23
10	250	51	23	250	51	23
20	250	47	24	250	51	22
30	250	26	24	250	42	23
60	250	26	23	250	21	22

Further, the performance is compared on two other parameters i.e. the number of iterations elapsed in the training of the network and elapsed time in training. Table 5.3 shows the comparison of the number of iterations elapsed in convergence. The backpropagation algorithm-based framework consumed 250 iterations which is the maximum number of iterations, on the other hand, the frameworks based on SaDE and BaDE algorithms took a very less number of iterations. The convergence was achieved by the adaptive learning algorithms very quickly in terms of iterations. However, these algorithms took a longer time in the training due to the fact that these algorithms work around a set of solutions whereas the backpropagation algorithm works with a single solution as shown in Table 5.4. The results obtained by the adaptive learning algorithms are very promising and can be used for better data center management.

TABLE 5.4 Training time (sec) elapsed in the training of predictive models using backpropagation, self-adaptive differential evolution, and biphase adaptive differential evolution

PWS (min)	NASA Trace			Saskatchewan Trace		
	WP_{BPNN}	WP_{SaDE}	WP_{BaDE}	WP_{BPNN}	WP_{SaDE}	WP_{BaDE}
1	121.368	290.823	771.250	661.690	1949.648	14506.710
5	24.112	48.539	237.210	132.666	498.452	1566.094
10	12.144	31.700	145.464	64.807	168.668	718.919
20	6.354	14.759	69.989	33.376	85.497	339.443
30	4.274	5.507	46.285	21.429	46.741	245.060
60	2.908	3.502	23.183	10.733	12.238	236.655

This chapter discussed two learning algorithms by extending one of the differential evolution variants. The learning algorithms (SaDE and BaDE) are used to train a neural network for workload forecasting. These methods are compared with state-of-art prediction models over two real-world benchmark data traces. It was observed that the model trained by the SaDE learning algorithm produced better forecasts in comparison to the state-of-art methods. The forecast accuracy of SaDE based prediction model was further enhanced by a model trained by BaDE algorithm. Thus, adaption at different levels helps in achieving better accuracy or improving the learning process.

CHAPTER 6

Self Directed Learning

$\mathbf{N}$ EURAL NETWORKS are the computational machines that try to mimic the learning behavior of a biological brain. Learning by the experience is one of the most common and widely used learning approaches in which a network learns by the means of feedback on its previous learning attempts. In this approach of learning, the network is unable to learn the changes in the data patterns that appear regularly in dynamic data sets such as cloud workloads. The cloud workloads are dynamic and may notice frequent changes in the data patterns. In such scenarios, a predictive framework requires frequent training over a period of time and thus, the model may become unreliable over a long period. In this chapter, we will discuss an alternative approach that avoids the need of frequent training.

The alternative learning approach allows a system to receive the feedback consistently so that it can capture the changes in the workload patterns which appear over time. The alternative approach is referred to as self-directed learning and the blackhole optimization algorithm is selected to demonstrate the impact of self-directed learning over a typical learning algorithm. The model analyses the most recent workload patterns to anticipate the load on the server in the next instance. It computes the average deviation in previous l predictions and leverages it in computing the next forecast which can be further utilized by the data center resource manager.

6.1   NON-DIRECTED LEARNING-BASED FRAMEWORK

The non-directed predictive framework (WP$_{\mathrm{BhNN}}$) is composed of a multi-layer neural network that analyzes the recent n workload instances to anticipate the next value. The framework design is depicted in Fig. 6.1, where the network employs the BhOA to learn its synaptic connection weights. The blackhole optimization algorithm is a population-based search optimization algorithm that follows the principles of the blackhole phenomenon. The next section discusses the algorithm in brief and interested readers are suggested to follow [57] for more details.

DOI: 10.1201/9781003110101-6

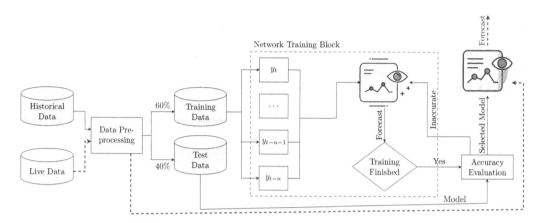

FIGURE 6.1 Non-directed predictive framework

6.1.1 Non-Directed Learning

A multi-layered neural network that consists of three layers i.e. input, hidden, and output layers is employed as the underlying architecture of the model. The neural network is composed of n, p, and q neurons in input, hidden, and output layers respectively. Let $X = \{x_1, x_2, \ldots, x_n\}$ be the n workload instances that act as an input to the forecaster. The input and output layer neurons use a linear function to process the information which they receive as input while hidden layer neurons process the information using a sigmoid function as shown in eq. (6.1).

$$f(z) = \begin{cases} z & \text{if input layer} \\ \frac{1}{1+e^{-z}} & \text{otherwise} \end{cases} \tag{6.1}$$

Similar to any population-based search algorithm, the blackhole algorithm also begins with a pool of randomly generated solutions in the search space using eq. (6.2). The search space is defined by the combination of synaptic network connections in (-1, 1). These solutions are further evaluated on training data samples and forecast error is measured using mean squared error. Since the optimal forecasting model must produce no error in the estimations, the aim of the network is to learn the synaptic weights that produce the minimum error in forecasts as given in eq. (6.3).

$$s_{(i,j)} = lb_j + r \times (ub_j - lb_j) \tag{6.2}$$

$$f_{cost} = min(\frac{1}{m} \sum_{i=1}^{m} (\hat{x}_i - x_i)^2) \tag{6.3}$$

Every individual solution (commonly referred to as a star) is assigned with its fitness value (root mean squared error). The stars are scattered all around the search space and the best star (one with the lowest root mean squared error) is called as blackhole ($\mathcal{B}$). Non-blackhole stars move towards the blackhole and their position gets updated as shown in eq. (6.4), where s_i depicts the position of ith star, r represents the random number in $[0, 1]$, and $\mathcal{B}$ depicts the position of the blackhole.

$$s_i(t+1) = s_i(t) + r \times (\mathcal{B}(t) - s_i(t)) \qquad (6.4)$$

$$\rho = \frac{f_{\mathcal{B}}}{\sum_{i=1}^{N} f_{s_i}} \qquad (6.5)$$

$$d_i = f_{\mathcal{B}} - f_{s_i} \qquad (6.6)$$

During the search process, a star $(s_i, i = 1, 2, \ldots, N)$ may find the better solution than $\mathcal{B}$. In such cases, the blackhole solution gets updated by exchanging the positions of s_i and $\mathcal{B}$. A star may reach into the event horizon radius (ρ) of the blackhole solution. The event horizon is defined by the radius computed using eq. 6.5, where the fitness values of $\mathcal{B}$ and s_i are defined by $f_{\mathcal{B}}$ and f_{s_i}. In order to check if a star has reached the event horizon, the distance between the star and blackhole is calculated as given in eq. 6.6. If the distance between s_i and $\mathcal{B}$ is less than or equal to ρ, the s_i is collapsed and a new star is generated randomly to maintain the size of the pool of stars.

6.2   SELF-DIRECTED LEARNING-BASED FRAMEWORK

As discussed in the previous section, a typical learning algorithm that only learns in the training phase is referred to as a non-directed learning algorithm. As opposed to these algorithms, if an algorithm also learns during the testing and deployment phases of the algorithm then it is referred to as a self-directed learning algorithm. This section discusses a predictive framework that uses a modified (self-directed) learning algorithm. The modified algorithm incorporates the principles of error preventive scheme i.e. it keeps track of the error in the recent forecasts and fines tune the predictions accordingly. Figure 6.2 depicts the complete process flow of the predictive framework equipped with a self-directed learning algorithm, where the feedback of the forecast errors is propagated back and an average of deviations is observed and incorporated in the next predictions. In the subsequent sections, the framework is discussed in the detail.

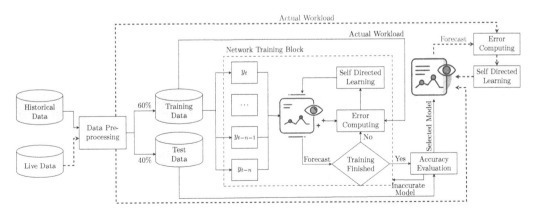

FIGURE 6.2 Self-directed predictive framework [82]

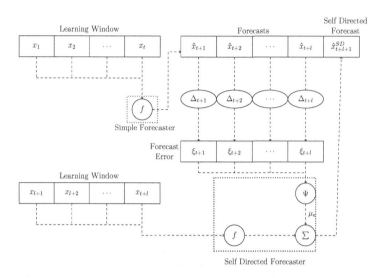

FIGURE 6.3 Self-directed learning process [82]

6.2.1 Self-Directed Learning

The self-directed learning algorithm learns the trend in the most recent forecasts i.e. a value is calculated which indicates the error in the forecast. This value is used to fine-tune the upcoming forecasts. The algorithm applies the Δ operator (eq. 6.7) to learn the forecast errors and uses Ψ operator over l recent prediction errors as shown in eq. 6.8. The basic functioning of both non-directed and self-directed learning is shown in Fig. 6.3.

$$\xi_{t+1} = \Delta_{t+1}(x_{t+1}, \hat{x}_{t+1}) \tag{6.7}$$
$$= x_{t+1} - \hat{x}_{t+1}$$
$$\mu_e = \Psi(\Delta_1, \Delta_2, \dots, \Delta_l) \tag{6.8}$$
$$= \frac{1}{l} \sum_{i=1}^{i=l} \Delta_i(x_i, \hat{x}_i)$$

For workload instances $X = \{0.15, 0.97, 0.95, 0.48, 0.80, 0.14, 0.42, 0.91, 0.79, 0.95\}$ an arbitrary forecaster predicts $\hat{X} = \{0.0, 1.17, 1.17, 0.83, 1.06, 0.59, 0.79, 1.14, 1.05, 1.17\}$ and its corresponding self directed forecaster predicts $\hat{X}^{SD} = \{0.0, 1.17, 1.17, 0.75, 0.83, 0.42, 0.60, 0.97, 0.87, 1.06\}$ by incorporating the error in previous three forecasts. The mean squared error for both directed and non-directed forecasts are 0.03 and 0.08 respectively. Thus, the incorporation of feedback from recent forecasts helps in improving the quality of model.

6.2.2 Cluster-Based Learning

The blackhole optimization algorithm is one of the simplest population-based search algorithms as it is parameter less i.e. the algorithm does not use any control parameters

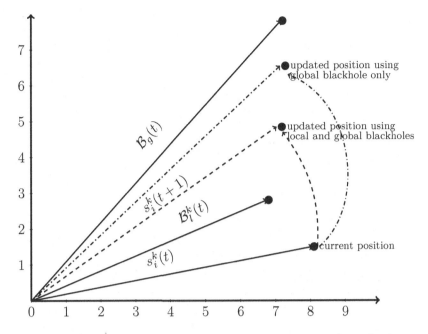

FIGURE 6.4 Position update procedures in blackhole algorithm (standard vs. modified)

such as mutation rate and crossover rate. The simplified search process guided by the best solution only may often lead towards the worst solution. For instance, if the best solution in the initial period of search gets stuck in local optima, the other solutions would also start moving towards the current best solution i.e. $\mathcal{B}$. Thus, there is a high probability of getting the algorithm trapped into local optima. A modified blackhole algorithm is discussed in [82], where the population is organized into a group of clusters. The algorithm avoids premature convergence by the means of dividing the populations into multiple subgroups and each subgroup explores the search space with their respective blackhole solutions. Let $S = \{s_1, s_2, \ldots, s_N\}$ be the pool of randomly initialized solutions (eq. (6.2)) arranged into $k = \{1, 2, \ldots, c\}$ different clusters (commonly referred to as subpopulations). Every sub-population contains N/c members and every member gets assigned the fitness value. Later, every cluster marks its respective blackhole ($\mathcal{B}_l^k$) and the best solution across the clusters becomes the global best (blackhole) solution ($\mathcal{B}_g$). In order to update the position of a solution s_i^k, it gets attracted by the $\mathcal{B}_l^k$ and $\mathcal{B}_g$ as shown in eq. (6.9), where $s_i^k(t)$ and $s_i^k(t+1)$ indicate the positions of an individual at time t and $t+1$, r_1 and r_2 represent two distinct random numbers in (0, 1), and α_l^k and α_g represent the forces applied by local and global blackhole solutions respectively. The division of the population members into multiple subpopulations and inclusion of local and global best information in the star movements slows down the movement of a solution towards the global best solution as shown in Fig. 6.4. Thus, the probability of premature convergence is reduced significantly.

$$s_i^k(t+1) = s_i^k(t) + \alpha_l^k \times r_1 \times (\mathcal{B}_l^k(t) - s_i^k(t)) + \alpha_g \times r_2 \times (\mathcal{B}_g(t) - s_i^k(t)) \quad (6.9)$$

$$r_h(\mathcal{B}_l^k) = f_{\mathcal{B}_l^k} / \sum_{i=1}^{N/c} f_{s_i^k} \tag{6.10}$$

$$r_h(\mathcal{B}_g) = f_{\mathcal{B}_g} / \sum_{k=1}^{c} \sum_{i=1}^{N/c} f_{s_i^k} \tag{6.11}$$

$$d_{\mathcal{B}_l}(s_i^k) = f_{\mathcal{B}_l^k} - f_{s_i^k} \tag{6.12}$$

$$d_{\mathcal{B}_g}(s_i^k) = f_{\mathcal{B}_g} - f_{s_i^k} \tag{6.13}$$

During the search process, if any sub-population member finds a better solution than its blackhole solution then it becomes the updated blackhole. Moreover, if its fitness value is better than the global blackhole then the global blackhole is also updated. Further, any star gets collapsed if it enters into the event horizon of a blackhole solution and a blackhole computes its radius using eqs. (6.10) and (6.11) (local and global) to mark the area for event horizon. A star enters into the event horizon of a blackhole if the distance (see eqs. (6.12) and (6.13)) between blackhole is less than or equal to the radius ($r_h(\mathcal{B}_l^k)$ or $r_h(\mathcal{B}_g)$) of the event horizon area. The rest of the search process is the same as the standard blackhole algorithm.

6.2.3 Complexity analysis

The non-directed and self-direct learning algorithms are analyzed on time complexity. Algorithm 6.1 presents the sequence of pseudo steps to be followed in a non-directed learning-based predictive framework. Similar to any population-based algorithms, the initialization step consumes $\mathcal{O}(n^2 N)$ to generate N solutions of length D. Similarly the fitness assessment step consumes $\mathcal{O}(mn^2 N)$ as discussed in the previous chapters. Further, line 3 selects the best solution with the lowest forecast error and designates it to be the blackhole of the population. It may consume $N-1$ comparisons to find out the best element therefore it becomes $\mathcal{O}(N)$. Next, an iterative process is repeated to find out the best possible network to forecast the future workload. For instance, line 6 updates the position of each solution except the $\mathcal{B}$ that needs $\mathcal{O}(D)$ to update the single solution position. Line 7 evaluates the updated solutions on f_{cost} and requires $\mathcal{O}(mn^2)$ for each network. Then $\mathcal{B}$ is updated if any of the newly generated solutions found better weights (line 8- 10) that requires $\mathcal{O}(D)$ to compare and update the blackhole with one solution. The total complexity of lines 5-11 becomes $\mathcal{O}(Nmn^2)$. Line 12 computes the radius of blackhole solution that needs $\mathcal{O}(N)$ as it computes the sum of the fitness values of each solution. Lines 13-18 compute the distance of each s_i from BhNN and s_i gets collapsed and regenerated if crosses the radius horizon of the blackhole that requires $\mathcal{O}(Nn^2)$. The total complexity of lines 4-19 becomes $\mathcal{O}(GNmn^2)$ as the process is repeated for G times. After summing up all the complexities, the time complexity of the approach was observed to be $\mathcal{O}(GNmn^2)$.

Algorithm 6.1 Pseudocode of WP_{BhNN} predictive model [73]

1: Randomly initialize a set (S) of N networks
2: Evaluate each network on objective function f_{cost}
3: Select the best solution to be black hole $(\mathcal{B})$ of population
4: **while** Termination criteria is not met **do**
5:     **for** each network $s_i(i \neq \mathcal{B}.Idx)$ **do**
6:         Update positions using eq. (6.4)
7:         Evaluate updated network using f_{cost}
8:         **if** $f_{s_i} < f_{\mathcal{B}}$ **then**
9:             Interchange position of $\mathcal{B}$ and s_i
10:         **end if**
11:     **end for**
12:     Calculate horizon radius ρ
13:     **for** each $s_i(i \neq \mathcal{B}.Idx)$ **do**
14:         Calculate distance d_i
15:         **if** $d_i \leq \rho$ **then**
16:             Collapse s_i and generate a new network
17:         **end if**
18:     **end for**
19: **end while**

The self-directed learning algorithm (Algorithm 6.2) organizes the population into c clusters and finds a blackhole for each cluster. The major updates in the algorithm are in the process of position updates. The additional costs in the algorithm are $\mathcal{O}(N)$ to determine c local blackholes, $\mathcal{O}(c)$ to determine global blackhole, and a constant time in maintaining c best solutions. Furthermore, the position update procedure also adds a constant amount of data. Thus, the modified algorithm consumes time in a similar order as a non-directed learning algorithm consumes [82].

6.3 FORECAST ASSESSMENT

An experimental analysis on D_1, D_2, D_3, D_4, D_5, and D_6 is conducted to assess the performance of both frameworks. The experimental environment is set using $n = 10$, $p = 7$, $G = 250, N = 100$, $c = 4$, $\alpha_l^k = 0.3$ and $\alpha_g = 0.8$ [82]. The model uses 60% of data traces to estimate the model parameters i.e. synaptic connection weights.

6.3.1 Short Term Forecast

The data traces belong to the different categories of workload i.e. the number of web requests on web servers, computing resource demands on cloud servers, and resource utilization on cloud servers. Therefore, the performance assessment is categorized into two different categories i.e. web server workloads and cloud server workloads.

Algorithm 6.2 Operational Summary of $\text{WP}_{\text{BhNN}}^{\text{SDL}}$ Method [82]

1: randomly initialize $S = \{s_i, s_2, \ldots, s_N\}$
2: organize S into c clusters and evaluate each s_i^k on training data
3: **for** $k = 1, 2, \ldots, c$ **do**
4:    $\mathcal{B}_l^k = Best(s_1^k, s_2^k, \ldots, s_{N/c}^k)$
5: **end for**
6: $\mathcal{B}_g = Best(\mathcal{B}_l^1, \mathcal{B}_l^2, \ldots, \mathcal{B}_l^c)$
7: **while** Termination criteria is not met **do**
8:    update position of each s_i^k using (6.9)
9:    evaluate updated stars $S(t+1)$ on training data
10:    **for** $k = 1, 2, \ldots, c$ **do**
11:       $\mathcal{B}_l^k(t+1) = Best(\mathcal{B}_l^k(t), s_1^k, s_2^k, \ldots, s_{N/c}^k)$
12:    **end for**
13:    $\mathcal{B}_g(t+1) = Best(\mathcal{B}_g(t), \mathcal{B}_l^1, \mathcal{B}_l^2, \ldots, \mathcal{B}_l^c)$
14:    compute the radius and distances using (6.10) to (6.13)
15:    **for** $i = 1, 2, \ldots, N$ **do**
16:       **if** $(d_{\mathcal{B}_l}(s_i^k) \leq \mathcal{B}_l^k) \parallel (d_{\mathcal{B}_g}(s_i^k) \leq \mathcal{B}_g)$ **then**
17:          collapse s_i^k and regenerate using (6.2)
18:       **end if**
19:    **end for**
20: **end while**

6.3.1.1 Web Server Workloads

The forecast results obtained on D_1, D_2, and D_3 using a self-directed learning algorithm are depicted in Fig. 6.5. The model captured the workload pattern and it predicts the next forecast with reasonable accuracy. The auto-correlation of corresponding forecast errors is shown in Fig. 6.6, which shows that the forecast errors are random and it is very difficult to capture the random noise in the data traces. The mean squared error of both predictive frameworks is compared and the corresponding results are shown in Fig. 6.7 and it shows that the incorporation of self-directed learning improves the prediction ability of a model.

A relative improvement up to 96.33%, 105.10%, 183.02%, and 65.18% was observed corresponding to 1, 5, 10, and 20-minute window size on D_1. Similarly, 123.33%, 30.22%, 148.32%, and 69.38% on D_2 while 78.10% , 76.76%, 130.29%, and 258.54% on D_3 for 1, 5, 10, and 20-minute window size respectively. The forecast accuracy drops down as the length of the prediction window increases. The time elapsed in network training was also compared and found that WP_{BhNN} consumes less time as shown in Table 6.8 due to the fact that it has to maintain the global best only.

6.3.1.2 Cloud Workloads

Similar to the web server workloads, the performance of forecasting approaches on cloud resource workloads was analyzed. Figure 6.9 depicts the actual and estimated workload information for a 5-minute prediction window. Figures 6.9a, 6.9b, and 6.9c corresponds to the forecast results of CPU Trace, Memory Trace, and PlanetLab

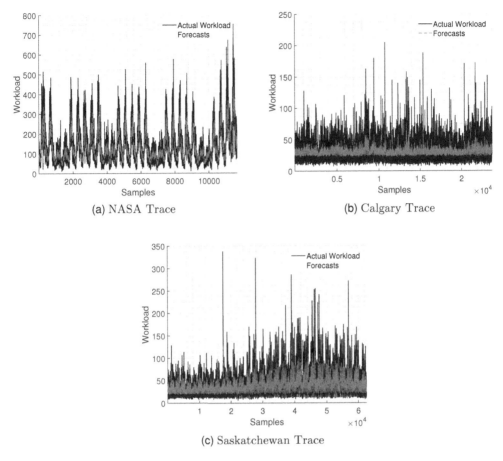

(a) NASA Trace

(b) Calgary Trace

(c) Saskatchewan Trace

FIGURE 6.5 Web server workload prediction results of self-directed learning predictive framework on 5-minute prediction interval [82]

Trace respectively. It can be observed that the forecasting model captures the trend and forecasts the future workload accordingly that can be verified from Fig. 6.10. It shows the auto-correlation in forecast residuals for each data-trace. The significant autocorrelation is not present in the forecast error except for few lags which validates that the forecast model captures historical data and extracts the meaningful pattern from it effectively.

The forecast accuracy measured using mean squared error is listed in Table 6.1 for both prediction schemes. A relative improvement was observed up to 80.16%, 84.03%, 99.44%, and 63.54% corresponding to 1, 5, 10, and 20-minute window size on D_4. Similarly, 13.24%, 104.58%, 151.03%, and 96.82% on D_5 for 1, 5, 10, and 20-minute window size respectively. The forecast accuracy drops down as the length of the prediction interval increases due to the fact that the number of training samples is reduced. The time elapsed in network training is also compared and it is found that WP$_{BhNN}$ consumes less time as shown in Table 6.2.

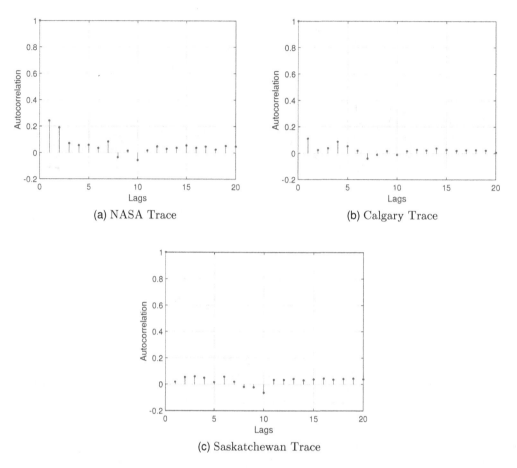

(a) NASA Trace

(b) Calgary Trace

(c) Saskatchewan Trace

FIGURE 6.6 Web server workload prediction residuals auto-correlation of self-directed learning predictive framework on 5-minute prediction interval [82]

6.4 LONG TERM FORECAST

In this section, the forecast accuracy of designed prediction approaches is evaluated on a long-term window. The respective experiments are carried out on the workload forecasting over different data traces (D_1, D_2, D_3, D_4, D_5, and D_6). The experiment setting involves the same parameters as of short-term forecast evaluation.

6.4.0.1 Web Server Workloads

Figure 6.11 depicts the forecasts and their corresponding actual values for 60-minute window. It was observed that the model is capable of detecting the trend of incoming workload on the server i.e. number of HTTP requests per time unit. The model gives an estimation of future load information on the servers as per the extracted pattern from the training data. Figure 6.12 shows the correlation in forecast error for respective traces and it can be observed that the forecast errors have some correlation due to the presence of spikes in the workloads.

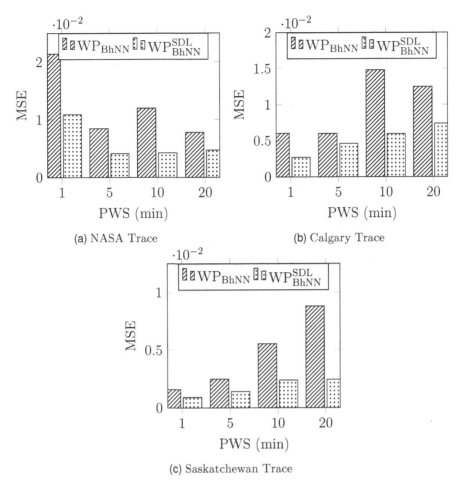

FIGURE 6.7 Mean squared error of non-directed and self-directed predictive frameworks on short term forecasts of web server workloads

The forecast accuracy of the models is given in Fig 6.13 that again validates the improvement in the modified approach of learning. A relative improvement of 79.05% and 169.54% corresponding to 30 and 60-minute window size on D_1 was observed. Similarly, 72.82% and 74.34% on D_2 while 225.10% and 104.40% on D_3 for 30 and 60-minute window size respectively. It is noticed that the forecast accuracy is usually decreasing as the length of the prediction window increases because the number of training samples is reduced accordingly. The time elapsed in network training was also compared and found that WP_{BhNN} consumes less time as shown in Fig. 6.14.

6.4.0.2 Cloud Workloads

The forecast accuracy of the proposed schemes on traces of cloud servers traces for a prediction window length of 30 minutes and above was also analyzed. The forecast results of D_4, D_5, and D_6 are depicted in Figs. 6.15a, 6.15b, and 6.15c respectively. The results show that the forecast model learns the pattern from data

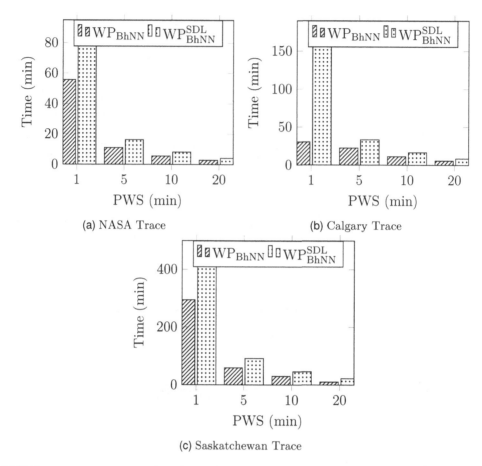

FIGURE 6.8 Training time (min) of non-directed and self-directed predictive frameworks on short term forecasts of web server workloads

and provides estimations close to actual workload information. The presence of noise in the forecasts is modeled using the autocorrelation in forecast residuals and it is evident from Fig. 6.16 that the forecasts are noisy due to the fact of the presence of spikes in the workload information. The forecast accuracy of the proposed approaches is listed in Table 6.3.

The study observed a relative improvement of 107.34% and 33.64% corresponding to 30 and 60-minute window size on D_4. Similarly, 62.43% and 0.93% on D_5 for 30 and 60-minute window size respectively. It was noticed that the forecast accuracy decreases as the length of the prediction window increases. The network training time is also compared and WP_{BhNN} consumes less time as compare to WP_{BhNN}^{SDL} as shown in Table 6.4.

6.5 COMPARATIVE & STATISTICAL ANALYSIS

This section compares the performance of both predictive frameworks with state-of-the-art approaches followed by statistical analysis on comparisons. Deep learning,

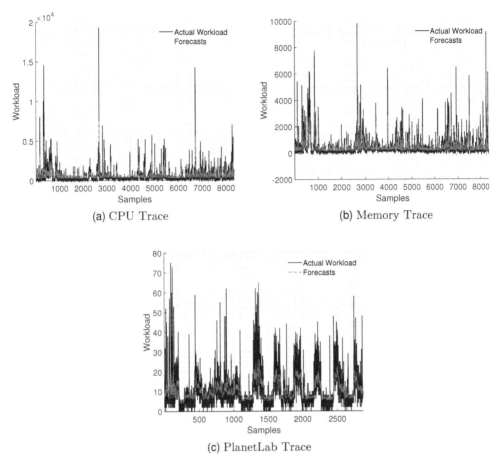

(a) CPU Trace

(b) Memory Trace

(c) PlanetLab Trace

FIGURE 6.9 Cloud server workload prediction results of self-directed learning predictive framework on 5-minute prediction interval [82]

differential evolution, and backpropagation-based prediction models are used to conduct the comparative study. In general, deep learning can be thought of as an extension of machine learning. The term deep refers to the usage of multiple hidden layers in the network. A number of predictive frameworks are developed using different architectures of deep neural networks such as recurrent neural networks. The performance of the self-directed learning-based algorithm is compared with [71, 137]. Differential Evolution is a numerical optimizer developed by Storn and Price [105]. It encodes the solutions in the form of vectors and manipulates them to explore the search space for a better solution. A large number of variants of the algorithms are available and one such popular variant is used for the purpose of the comparison [74]. As opposed to the differential evolution which is a population-based search algorithm, the backpropagation uses a single solution in the search and finds a better solution by propagating the error feedback which is computed using the concept of gradient [104].

Tables 6.5-6.10 list out the forecast accuracy and compare them for every datatrace. The results observed the superiority in the performance of self-directed learning-

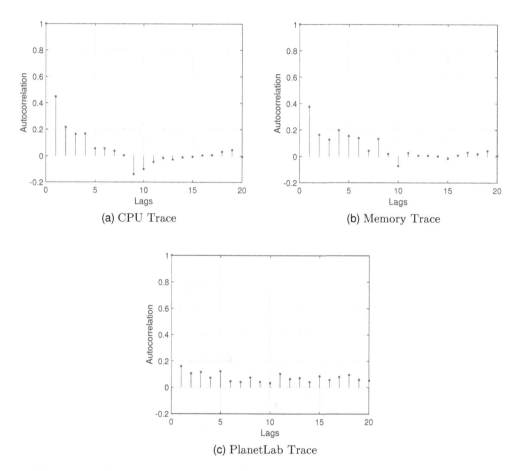

FIGURE 6.10 Cloud server workload prediction residuals auto-correlation of self-directed learning predictive framework on 5-minute prediction interval [82]

TABLE 6.1 Mean squared error of non-directed and self-directed predictive frameworks on short term forecasts of cloud server workloads

PWS (min)	WP_{BhNN}		WP^{SDL}_{BhNN}		
	D_4	D_5	D_4	D_5	D_6
1	9.260E-04	2.480E-04	5.140E-04	2.190E-04	-
5	2.190E-03	4.910E-03	1.190E-03	2.400E-03	1.040E-02
10	3.590E-03	8.560E-03	1.800E-03	3.410E-03	1.210E-02
20	6.280E-03	1.360E-02	3.840E-03	6.910E-03	1.310E-02

based algorithms. For better understanding, the error reduction ratio is computed using $MSE_{RED} = {}^{MSE_{OLD} - MSE_{NEW}}/{}_{MSE_{OLD}} * 100$, where MSE_{RED}, MSE_{NEW}, and MSE_{OLD} indicate the percentage reduction between the MSE of new and old (state of art) approaches. For instance, on NASA trace the maximum relative reduction

TABLE 6.2 Training time (min) of non-directed and self-directed predictive frameworks on short term forecasts of cloud server workloads

PWS (min)	WP$_{\text{BhNN}}$		WP$_{\text{BhNN}}^{\text{SDL}}$		
	D_4	D_5	D_4	D_5	D_6
1	41.86	40.84	60.61	60.3	NA
5	1.10	8.11	11.86	12.1	4.11
10	0.55	3.94	6.03	5.87	2.1
20	2.10	0.22	2.99	2.96	1.05

is up to $36.340\%, 64.670\%, 83.040\%$, and 98.640% over deep learning, non-directed blackhole learning, self-adaptive differential evolution, and backpropagation based models. Similarly, the Calgary trace forecasts notice the relative reduction up to $21.050\%, 59.730\%, 9.450\%$, and 99.090%, and the Saskatchewan trace notice reduction

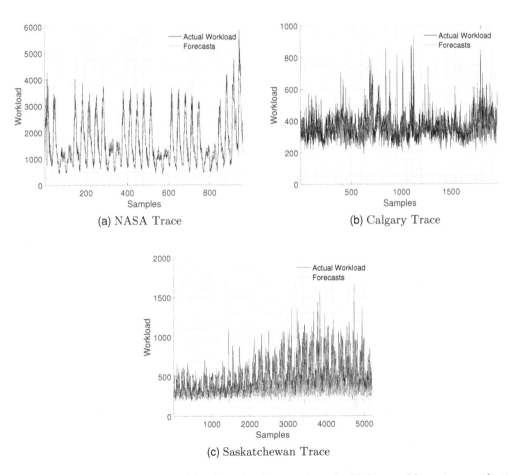

(a) NASA Trace

(b) Calgary Trace

(c) Saskatchewan Trace

FIGURE 6.11 Web server workload prediction results of self-directed learning predictive framework on 60-minute prediction interval [82]

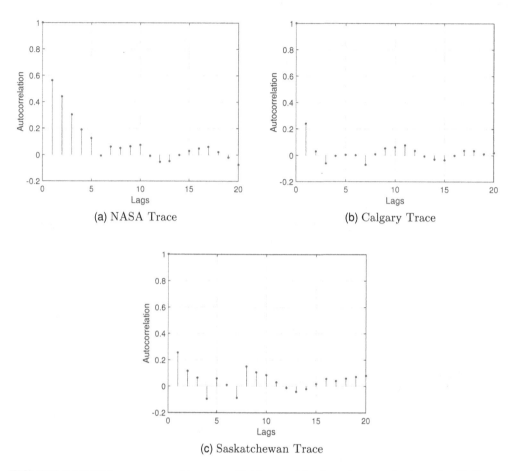

(a) NASA Trace

(b) Calgary Trace

(c) Saskatchewan Trace

FIGURE 6.12 Web server workload prediction residuals auto-correlation of self-directed learning predictive framework on 60-minute prediction interval [82]

up to $81.920\%, 72.110\%, 94.350\%$, and 99.580% over state of art models. On cloud workloads, the maximum reduction in the error is noticed up to $51.770\%, 39.390\%$, and 89.610% for CPU trace while $60.160\%, 25.550\%$, and 88.880% for a Memory trace. Furthermore, the error reduction up to 99.990% is observed on PlanetLab cpu utilization trace forecasts. These results are further statistically analyzed to ensure the correctness of the results.

The statistical test is conducted using the Friedman test [41] which conducts multiple test comparisons to find out the significant difference in the results. It computes a rank for each result to compare the forecast quality and corresponding results are shown in Table 6.11. The self-directed learning-based predictive model gets the better rank among these models. However, the deep learning-based model gets a better rank on the forecast of Calgary trace. Moreover, the Friedman test statistics (χ^2 and p-value) are shown in Table 6.12, where it can be seen that the null hypothesis of the test is rejected for $\aleph = 0.05$ meaning that the results are statistically different.

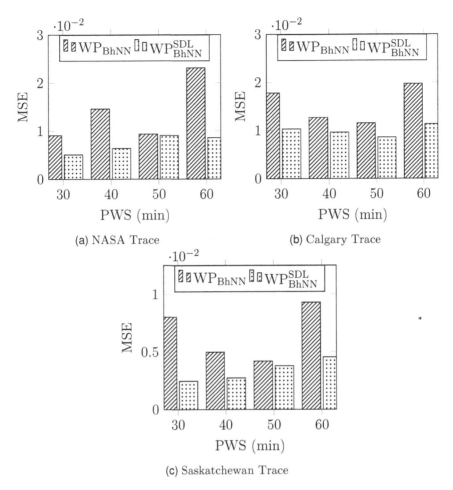

(a) NASA Trace

(b) Calgary Trace

(c) Saskatchewan Trace

FIGURE 6.13 Mean squared error of non-directed and self-directed predictive frameworks on long term forecasts of web server workloads

Unfortunately, the Friedman test is unable to identify the significant difference between the two results. The Wilcoxon signed-rank test helps in checking if the two results are significantly different or not. This test performs a series of pairwise tests and calculates the positive and negative mean rankings which are denoted as R_{WC}^{+} and R_{WC}^{-} respectively. The corresponding results are listed in Table 6.13, where - (hyphen) indicates that either value is not computed or is not available. The self-directed learning-based predictive framework achieves better results over its competitors since the test rejects the null hypothesis on a significance level of 0.05.

In this chapter, we have discussed the self-directed learning-based predictive model which learns from its previous mistakes. The underlying learning algorithm is blackhole optimization. The advantage of the learning algorithm is that its performance is independent of any internal parameter optimization such as crossover rate and mutation rate as opposed to other population-based optimization algorithms. A modified learning algorithm is also discussed which organizes the population into

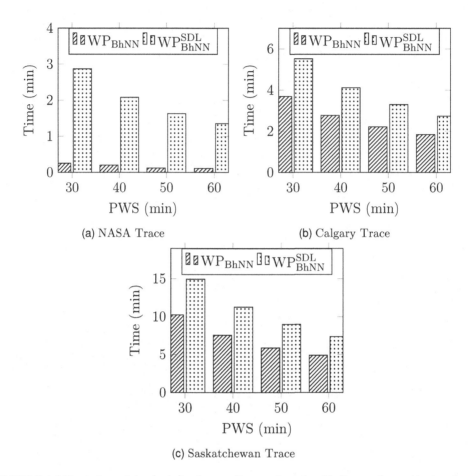

(a) NASA Trace   (b) Calgary Trace

(c) Saskatchewan Trace

FIGURE 6.14 Training time (min) of non-directed and self-directed predictive frameworks on long term forecasts of web server workloads

a set of clusters and allows to incorporate the local and global best information to generate new solutions. Thus, the reorganization of the population and incorporation of self-directed learning improved the forecast quality significantly.

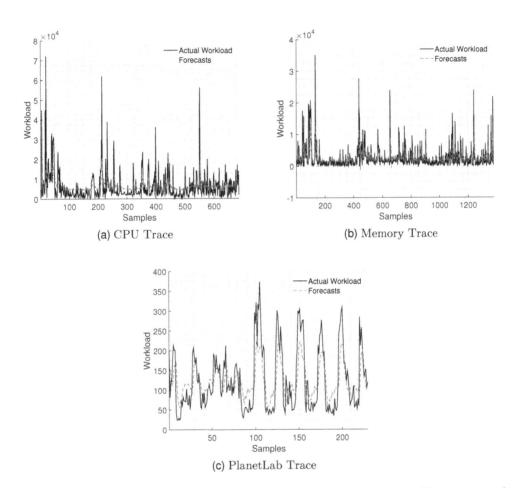

(a) CPU Trace

(b) Memory Trace

(c) PlanetLab Trace

FIGURE 6.15 Cloud server workload prediction results of self-directed learning predictive framework on 60-minute prediction interval [82]

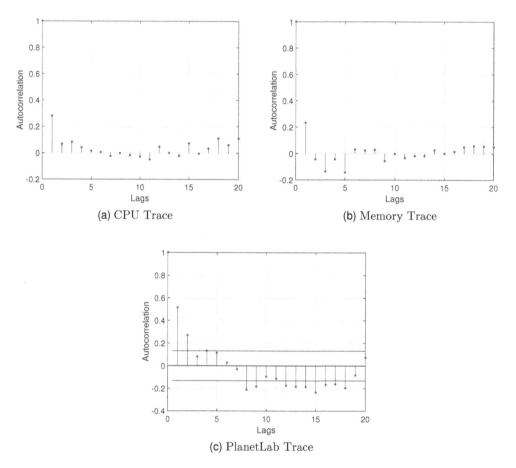

(a) CPU Trace

(b) Memory Trace

(c) PlanetLab Trace

FIGURE 6.16 Cloud server workload prediction residuals auto-correlation of self-directed learning predictive framework on 60-minute prediction interval [82]

TABLE 6.3 Mean squared error of non directed and self-directed predictive frameworks on long term forecasts of cloud server workloads

PWS	WP$_{BhNN}$		WP$_{BhNN}^{SDL}$		
(min)	D$_4$	D$_5$	D$_4$	D$_5$	D$_6$
30	1.130E-02	1.150E-02	5.450E-03	7.080E-03	1.430E-02
40	7.640E-03	1.460E-02	8.420E-03	8.760E-03	1.460E-02
50	1.210E-02	1.120E-02	8.480E-03	1.170E-02	1.640E-02
60	1.430E-02	1.090E-02	1.070E-02	1.080E-02	1.640E-02

TABLE 6.4 Training time (min) of non-directed and self-directed predictive frameworks on long term forecasts of cloud server workloads

PWS	WP$_{BhNN}$		WP$_{BhNN}^{SDL}$		
(min)	D$_4$	D$_5$	D$_4$	D$_5$	D$_6$
30	1.41	1.30	1.99	1.97	0.68
40	0.17	0.98	1.48	1.49	0.52
50	0.10	0.79	1.19	1.19	0.41
60	0.70	0.65	0.98	1.00	0.33

TABLE 6.5 Mean squared error comparison of non-directed and self-directed learning based models' NASA Trace forecasts with state-of-art models [82]

PWS (min)	WP$_{BhNN}^{SDL}$	LSTM	WP$_{BhNN}$	SaDE	BPNN
1	1.090E-02	1.310E-02	2.140E-02	1.690E-04	2.430E-01
5	4.120E-03	4.790E-03	8.450E-03	1.000E-02	3.020E-01
10	4.240E-03	6.660E-03	1.200E-02	2.500E-02	2.820E-01
20	4.710E-03	7.010E-03	7.780E-03	2.500E-02	3.380E-01
30	5.060E-03	6.430E-03	9.060E-03	2.020E-02	2.780E-01
60	8.570E-03	5.590E-03	2.310E-02	2.020E-02	3.340E-01

TABLE 6.6 Mean squared error comparison of non-directed and self-directed learning based models' Calgary Trace forecasts with state-of-art models [82]

PWS (min)	WP$_{BhNN}^{SDL}$	LSTM	WP$_{BhNN}$	SaDE	BPNN
1	2.700E-03	3.420E-03	6.030E-03	2.790E-03	2.970E-01
5	4.600E-03	4.100E-03	5.990E-03	5.080E-03	2.900E-01
10	5.960E-03	6.110E-03	1.480E-02	6.060E-03	2.870E-01
20	7.380E-03	5.990E-03	1.250E-02	7.790E-03	2.780E-01
30	1.030E-02	7.120E-03	1.780E-02	1.020E-02	5.000E-01
60	1.130E-02	8.030E-03	1.970E-02	1.170E-02	2.970E-01

TABLE 6.7 Mean squared error comparison of non-directed and self-directed learning based models' Saskatchewan Trace forecasts with state-of-art models [82]

PWS (min)	$\text{WP}_{\text{BhNN}}^{\text{SDL}}$	LSTM	WP_{BhNN}	SaDE	BPNN
1	9.040E-04	5.000E-03	1.610E-03	1.000E-06	4.020E-02
5	1.420E-03	3.170E-03	2.510E-03	4.000E-06	3.370E-01
10	2.410E-03	5.260E-03	5.550E-03	2.500E-02	2.900E-01
20	2.460E-03	5.560E-03	8.820E-03	3.840E-02	3.180E-01
30	2.470E-03	4.790E-03	8.030E-03	4.370E-02	5.070E-01
60	4.550E-03	4.500E-03	9.300E-03	2.890E-02	2.860E-01

TABLE 6.8 Mean squared error comparison of non-directed and self-directed learning based models' CPU Trace forecasts with state-of-art models [82]

PWS (min)	$\text{WP}_{\text{BhNN}}^{\text{SDL}}$	WP_{BhNN}	SaDE	BPNN
1	5.140E-04	9.260E-04	8.480E-04	4.410E-03
5	1.190E-03	2.190E-03	1.680E-03	8.750E-03
10	1.800E-03	3.590E-03	2.470E-03	1.420E-02
20	3.840E-03	6.280E-03	4.020E-03	2.600E-02
30	5.450E-03	1.130E-02	5.490E-03	4.370E-02
60	1.070E-02	1.430E-02	1.030E-02	1.030E-01

TABLE 6.9 Mean squared error comparison of non-directed and self-directed learning based models' Memory Trace forecasts with state-of-art models [82]

PWS (min)	$\text{WP}_{\text{BhNN}}^{\text{SDL}}$	WP_{BhNN}	SaDE	BPNN
1	2.190E-04	2.480E-04	2.350E-04	1.970E-03
5	2.400E-03	4.910E-03	2.230E-03	1.910E-02
10	3.410E-03	8.560E-03	4.080E-03	2.480E-02
20	6.910E-03	1.360E-02	5.540E-03	4.800E-02
30	7.080E-03	1.150E-02	9.510E-03	5.160E-02
60	1.080E-02	1.090E-02	1.180E-02	8.770E-02

TABLE 6.10 Mean squared error comparison of self-directed learning-based model's PlanetLab Trace forecasts with deep learning model [82]

PWS (min)	$\text{WP}_{\text{BhNN}}^{\text{SDL}}$	Deep Learning
5	1.040E-02	8.410E+01
15	1.330E-02	9.290E+01
30	1.430E-02	1.060E+02
60	1.640E-02	9.940E+01

TABLE 6.11 Friedman test ranks of non-directed, self-directed, LSTM, SaDE, and backpropagation based predictive models [82]

Prediction Model	D_1	D_2	D_3	D_4	D_5
WP_{BhNN}^{SDL}	1.33	1.83	1.50	1.17	1.33
LSTM	2.00	1.67	2.50	-	-
WP_{BhNN}	3.33	4.00	3.00	3.00	2.83
SaDE	3.33	2.50	3.00	1.83	1.83
BPNN	5.00	5.00	5.00	4.00	4.00

TABLE 6.12 Friedman test statistics of non-directed and self-directed learning predictive frameworks [82]

	D_1	D_2	D_3	D_4	D_5
χ^2	19.200	20.133	15.600	17.000	15.000
p	0.001	0.000	0.004	0.001	0.002

TABLE 6.13 Wilcoxon signed test ranks for self-directed learning predictive framework [82]

	WP_{BhNN}^{SDL}	D_1	D_2	D_3	D_4	D_5
	R_{WC}^-	6.00	17.00	1.00	-	-
LSTM	R_{WC}^+	15.00	4.00	20.00	-	-
	p-value	0.345	0.173	0.046	-	-
	R_{WC}^-	0.00	0.00	0.00	0.00	0.00
WP_{BhNN}	R_{WC}^+	21.00	21.00	21.00	21.00	21.00
	p-value	0.028	0.028	0.028	0.028	0.028
	R_{WC}^-	2.00	2.50	3.00	4.00	7.00
SaDE	R_{WC}^+	19.00	18.50	18.00	17.00	14.00
	p-value	0.075	0.093	0.116	0.173	0.463
	R_{WC}^-	0.00	0.00	0.00	0.00	0.00
BPNN	R_{WC}^+	21.00	21.00	21.00	21.00	21.00
	p-value	0.028	0.028	0.028	0.028	0.028

CHAPTER 7

Ensemble Learning

NSEMBLE LEARNING considers the opinion of many-fold models while making a decision in solving computationally intelligent problems. This chapter discusses two forecasting approaches developed using Extreme Learning Machine (ELM) algorithm which was introduced by Guang-Bin Huang [51].

7.1 EXTREME LEARNING MACHINE

An ELM is a network learning algorithm that learns the synaptic connection weights in a single step. The weights corresponding to the input-hidden layer are randomly initialized and hidden-output layer weights are obtained by computing the Moore-Penrose inverse. Whereas, an ELM network is a feed-forward neural network trained by an ELM algorithm for different tasks such as prediction, detection, classification, clustering, and many others. The term ELM is interchangeably used for the learning algorithm and network trained by an ELM. The extreme learning machine networks of large scale are claimed to be several hundred times faster than iterative learning schemes such as backpropagation due to the fact that they learn the network parameters in a single step as opposed to iterative learning processes.

Let ω and $\tilde{\omega}$ be the weight matrices corresponding to the input-hidden layer connections and hidden-output layer connections respectively. If m paired data samples $(\varkappa_j, \varsigma_j)$ are used for training given $\varkappa_j \in \mathbb{R}^n$ and $\varsigma_j \in \mathbb{R}$, mathematically ELM can be formulated as given in eq. (7.1), where $\omega_i \in \mathbb{R}^p$ represents the weights corresponding to the connections from input neurons to i^{th} hidden neuron, $b_i \in \mathbb{R}$ represents the weight corresponding to the connection from bias neuron to i^{th} hidden neuron, $\tilde{\omega}_{ik} \in \mathbb{R}$ represents the connection weight between i^{th} hidden neurons and k^{th} output neuron, dot $(\cdot)$ operator represents the inner product, ζ represents the activation function (commonly sigmoid function), and $\varsigma_j \in \mathbb{R}$ represents the output value of the network.

$$\sum_{k=1}^{q} \sum_{i=1}^{p} \tilde{\omega}_{ik} \times \zeta\left(\omega_i \cdot \varkappa_j + b_i\right) = \varsigma_j, \quad j = 1, \cdots, m \qquad (7.1)$$

DOI: 10.1201/9781003110101-7

7.2 WORKLOAD DECOMPOSITION PREDICTIVE FRAMEWORK

The prediction approach extracts the resource demand traces from raw data and aggregates them in per unit time interval. The difference operation is performed over the aggregated traces to reduce the non-linearity. The order of difference operation is determined using the ARIMA [60] process followed by rescaling of the traces. The preprocessed workload trace is decomposed into different components and distinct networks are trained for each component as shown in Fig. 7.1. The estimations of each network are combined to anticipate the future workload information.

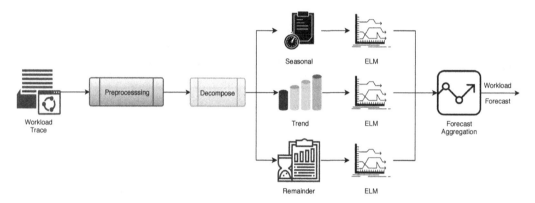

FIGURE 7.1 Decomposition based predictive framework

7.2.1 Framework Design

The workload trace is decomposed into three distinct components i.e. seasonal, trend, and remainder using seasonal decomposition [27], and seasonality is detected using Fourier transforms. Let x_t^s, x_t^t, and x_t^r be the seasonal, trend, and remainder components respectively corresponding to x_t. The sum of all these components define the actual workload i.e. $x_t = x_t^s + x_t^t + x_t^r$. Figure 7.2 renders the decomposition of CPU trace. Before decomposition, the trace is preprocessed using difference and min-max rescaling. The differential transformation can be explained as the difference between resource demand at time t and $t-1$. The value of difference order (d) is obtained by analyzing the workload trace using the ARIMA process. The workload trace is rescaled using eq. 7.2, where X_{norm}, X_{max}, and X_{min} are normalized, maximum and minimum values of workload trace respectively.

$$X_{norm} = \frac{X - X_{min}}{X_{max} - X_{min}} \tag{7.2}$$

Input to the predictor model is a sequence of n past and lagged resource demands. Each network of the prediction model is composed of a three-layered extreme learning machine. The single output node neural network can be interpreted as a non-linear function of input values. The accuracy of predictions made by extreme learning machines can be affected by numerous parameters such as the number of input nodes, hidden neurons, and the size of training data. The number of input nodes is

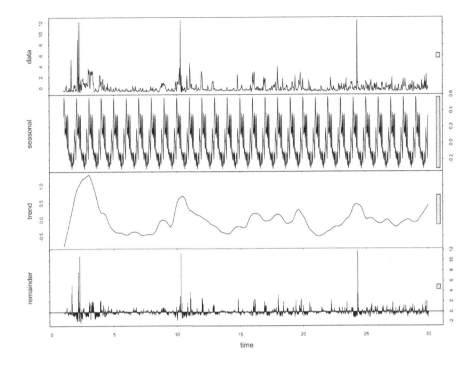

FIGURE 7.2 Decomposition of CPU requests trace [79]

approximated using auto ARIMA process. Table 7.1 lists the obtained parameters on original and differenced traces. Since the transformation order in each case is not more than 5, the input nodes must be close to 5. Thus, three different values 2, 5, and 10 are chosen for the number of input neurons. The selection of the number of hidden nodes is another critical issue in neural networks. Three different values for a number of hidden neurons are selected randomly. The details of opted values for all parameters are shown in Table 7.2. An experiment list is designed using these parameters' values through the D-Optimal Design method [32]. The prediction model is trained for each of the experiments listed in Table 7.3. After training, the machine is examined over unseen patterns. The mean prediction error (MPE) metric is used to evaluate the performance of the prediction model. Further predicted amounts of resources can be fed into the resource manager of the data center. These inputs can be utilized in the decision-making process regarding resource scaling. Thus SLA violations can be reduced by scaling resources in advance before actual demand arrives. An operational summary of the framework is given in Algorithm 7.1.

TABLE 7.1 ARIMA analysis orders for cloud resource demand traces [79]

Trace	Autoregression	Integration	Moving average
CPU Trace	3	1	5
CPU Trace (after first order difference)	3	0	5
Memory Trace	1	1	2
Memory Trace (after first order difference)	1	0	2

TABLE 7.2 Network configuration parameter choices for decomposition predictive framework [79]

Input Nodes	Hidden Nodes	Training Sample Size (%)
2	5	50
5	7	65
10	10	80

TABLE 7.3 List of experiments selected by D-Optimal Design [79]

Exp No	Input Node	Hidden Node	Training Data Size (%)
EXP-1	2	10	50
EXP-2	5	5	50
EXP-3	10	10	50
EXP-4	2	5	80
EXP-5	5	10	65
EXP-6	10	7	65
EXP-7	2	10	80
EXP-8	5	7	80
EXP-9	10	5	80
EXP-10	10	10	80

Algorithm 7.1 Pseudocode for WP_{ELMNN} predictive framework

1: Read workload trace (x)
2: Apply decomposition $[x^s, x^t, x^r] = \text{DECOMPOSITION}(x)$
3: Prepare input data according to n for all components
4: **for** each workload component $k = \{x^s, x^t, x^r\}$ **do**
5:    $\omega^{\varepsilon_k} = rand(n+1, p)$
6:    $\tilde{\omega}^{\varepsilon_k} = \hbar^{\varepsilon_k^\dagger} \boldsymbol{\tau}$
7: **end for**
8: /* Forecast evaluation on test data */
9: **for** each workload component $k = \{x^s, x^t, x^r\}$ **do**
10:    **for** each data sample $t = \{1, 2, \ldots, m\}$ **do**
11:       $\hat{x}_t^s = \varepsilon_{x^s}(x_{t-1}^s, x_{t-2}^s, \ldots, x_{t-n}^s)$
12:       $\hat{x}_t^t = \varepsilon_{x^t}(x_{t-1}^t, x_{t-2}^t, \ldots, x_{t-n}^t)$
13:       $\hat{x}_t^r = \varepsilon_{x^r}(x_{t-1}^r, x_{t-2}^r, \ldots, x_{t-n}^r)$
14:       $\hat{x}_t = \hat{x}_t^s + \hat{x}_t^t + \hat{x}_t^r$
15:       $\xi_t = x_t - \hat{x}_t$
16:    **end for**
17: **end for**

7.3 ELM ENSEMBLE PREDICTIVE FRAMEWORK

The single predictor is not always suitable for different types of workloads i.e. different predictors suit different workloads. The ensemble learning approach uses a combination of prediction models to calculate the final outcome. The ensemble model combines the forecast of the individual prediction model using a voting engine. Figure 7.3 shows the conceptual framework of a prediction model based on ensemble learning.

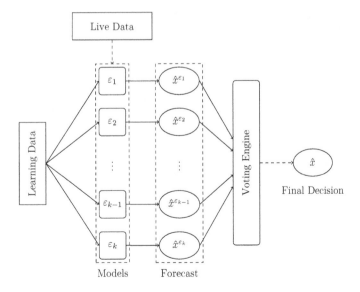

FIGURE 7.3 A conceptual view of ensemble stacking [81]

7.3.1 Ensemble Learning

The ensemble framework is composed of k multilayered neural networks or base experts. Every network consists of n, p, and q neurons arranged in input, hidden, and output layers respectively. The networks use $\zeta(\bullet)$ as an activation function at the hidden layer. Every base expert learns the synaptic connection weights using an extreme learning machine algorithm. An ELM selects the input to hidden layer connection weights randomly and hidden to output layer connection weights are calculated analytically [59]. The algorithm finds these connection weights by solving a general linear system. Moreover, the ELM algorithm is very popular for its speed and the ability of universal approximation [51].

Let $\varkappa = \{(\varkappa_j, \varsigma_j) | \varkappa_j \in \mathbb{R}^n, \varsigma_j \in \mathbb{R}\}$ be a set of workload patterns, every workload pattern is a combination of workload values of length of learning window. For instance, $\varkappa_j$ is a combination of $[x_{j-1}, x_{j-2}, \ldots, x_{j-n}]$ provided that the length of learning window in n. The data creation process is shown in eq. (7.3). Further, let $\omega_i^{\varepsilon_k} = [\omega_{1i}^{\varepsilon_k}, \omega_{2i}^{\varepsilon_k}, \ldots, \omega_{ni}^{\varepsilon_k}]$ be the synaptic connection weights between input neurons and i^{th} hidden neuron of ε_k and $b_i^{\varepsilon_k}$ be the connection weight of bias to $h_i^{\varepsilon_k}$ (i^{th} hidden neuron of ε_k). The input to the output neurons i.e. the activation of hidden neurons can be computed as $\hbar_i^{\varepsilon_k}(\varkappa_j) = \zeta(\omega_i^{\varepsilon_k} \cdot \varkappa_j + b_i^{\varepsilon_k})$, which is the output of $h_i^{\varepsilon_k}$ on $\varkappa_j$. Let $\tilde{\omega}_i^{\varepsilon_k} = [\tilde{\omega}_{i1}^{\varepsilon_k}, \tilde{\omega}_{i2}^{\varepsilon_k}, \ldots, \tilde{\omega}_{iq}^{\varepsilon_k}]^T$ be the set of synaptic connection weights between $h_i^{\varepsilon_k}$ and output neurons. Given the above-mentioned definitions, the forecast corresponding to j^{th} workload pattern can be obtained as shown in eq. (7.4).

$$\varkappa = \begin{bmatrix} x_1 & x_2 & \cdots & x_n \\ x_2 & x_3 & \cdots & x_{n+1} \\ \vdots & \vdots & \ddots & \vdots \\ x_m & x_{m+1} & \cdots & x_{n+m-1} \end{bmatrix} \begin{bmatrix} x_{n+1} \\ x_{n+2} \\ \vdots \\ x_{n+m} \end{bmatrix} \tag{7.3}$$

$$\hat{\varsigma}_j^{\varepsilon_k} = \sum_{i=1}^{p} \tilde{\omega}_i^{\varepsilon_k} \times \zeta(\omega_i^{\varepsilon_k} \cdot \varkappa_j + b_i^{\varepsilon_k}); \qquad \forall j \in \{1, 2, \ldots, m\} \tag{7.4}$$

Let $\hat{\varsigma}^{\varepsilon_k}$ be the set of forecast values generated by an expert ε_k which can be obtained using eq. (7.4), where $\sum_{i=1}^{p} \tilde{\omega}_i^{\varepsilon_k} \times \hbar_i^{\varepsilon_k}(\varkappa_j)$ can be written as $\tilde{\omega}^{\varepsilon_k} \times \hbar^{\varepsilon_k}$. Here, $\hbar^{\varepsilon_k}$ represents the hidden layer output and can be calculated using eq. (7.5). If a system can observe the values of ω^{ε_k}, $\tilde{\omega}^{\varepsilon_k}$, and b^{ε_k} such that $\hat{\varsigma}_j^{\varepsilon_k} = \varsigma_j$ for all m data patterns then the observed forecasts are generated with no error. Thus, the minimum squared prediction error can be calculated using $\min_{\tilde{\omega}^{\varepsilon_k}} \|\tilde{\omega}^{\varepsilon_k} \hbar^{\varepsilon_k} - \varsigma\|^2$.

$$\hbar^{\varepsilon_k} = \begin{bmatrix} \hbar^{\varepsilon_k}(\varkappa_1) \\ \hbar^{\varepsilon_k}(\varkappa_2) \\ \vdots \\ \hbar^{\varepsilon_k}(\varkappa_m) \end{bmatrix} = \begin{bmatrix} \hbar_1^{\varepsilon_k}(\varkappa_1) & \hbar_2^{\varepsilon_k}(\varkappa_1) & \cdots & \hbar_p^{\varepsilon_k}(\varkappa_1) \\ \hbar_1^{\varepsilon_k}(\varkappa_2) & \hbar_2^{\varepsilon_k}(\varkappa_2) & \cdots & \hbar_p^{\varepsilon_k}(\varkappa_2) \\ \vdots & \vdots & \ddots & \vdots \\ \hbar_1^{\varepsilon_k}(\varkappa_m) & \hbar_2^{\varepsilon_k}(\varkappa_m) & \cdots & \hbar_p^{\varepsilon_k}(\varkappa_m) \end{bmatrix} \tag{7.5}$$

An ELM algorithm solves a general linear system to learn the weights ($\tilde{\omega}^{\varepsilon_k}$) corresponding to the synaptic connections from hidden and output neurons of ε_k. The algorithm approximates $\tilde{\omega}^{\varepsilon_k}$ by randomly generating the weights from input and bias neurons to hidden neurons and then finds least-square solution for general linear system $\hbar^{\varepsilon_k} \cdot \tilde{\omega}^{\varepsilon_k} = \varsigma$ as shown in eq. (7.6), where $\tilde{\Omega}^{\varepsilon_k}$ and $\hbar^{\varepsilon_k^\dagger}$ are the least square solutions with minimum norm and uniqueness, and Moore-Penrose generalized inverse of a matrix $\hbar^{\varepsilon_k}$ respectively [81]. Every expert generates the forecasts after approximating the connection weights which are further weighted using a voting engine, where a heuristic-based approach assigns the weights (α^{ε_k}) for every expert forecast.

$$\tilde{\Omega}^{\varepsilon_k} = \hbar^{\varepsilon_k^\dagger}\varsigma \tag{7.6}$$

7.3.2 Expert Architecture Learning

The performance of a neural network highly depends on various parameters such as number of layers, number of neurons in each layer, connections between neurons, learning algorithm, etc. The ensemble-based predictive framework uses k neural networks which are composed of n, p, and q neurons arranged in input, hidden, and output layers as depicted in Fig. 7.4. The framework uses the networks having a fixed number of layers and the number of neurons in the output layer. Furthermore, the framework learns the number of neurons in input and hidden layers.

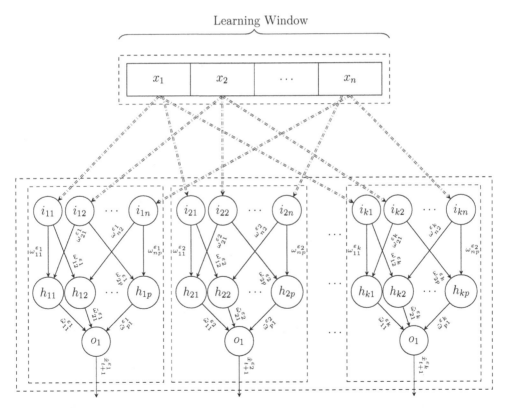

FIGURE 7.4 An ensemble of ELMs in workload prediction [81]

The number of input neurons is selected as per the data characteristics which means for data traces having different properties the framework may use a different number of input neurons. The framework obtains a suitable number of input neurons by modeling the auto-correlation of data-trace. The auto-correlation helps in finding the number of historical values that could be helpful in forecasting as it explains the internal association across the observations of a time series indexed data-trace. Let R_X be the auto-correlation of a data-trace, say X. First, the framework finds the auto-correlation for the first 40 time lags (L) as shown in eq. (7.7), where the term `autocorr` represents a function named autocorr which returns the auto-correlation for first L time lags. The term L is generic and any value can be chosen. Later, Υ is applied to the observed auto-correlation which returns the first lag with insignificant auto-correlation (eq. (7.8)), where T_ϱ defines the threshold to determine the significant auto-correlation. If there is no time lag obtained with insignificant threshold, the data trace is differentiated and the same process is repeated up to d (difference order) times. Otherwise, the obtained lag with insignificant threshold represents the number of input neurons in the network as shown in Fig. 7.5.

$$R_X = \texttt{autocorr}(X, L) \tag{7.7}$$

$$sl = \Upsilon((|R_X| \leq T_\varrho), 1) \tag{7.8}$$

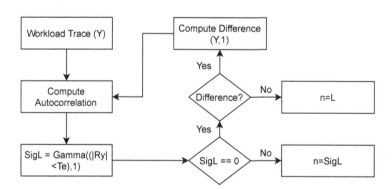

FIGURE 7.5 Input node selection

The framework also learns the suitable number of hidden neurons using three different heuristics (fix, linear, and random) as shown in eq. (7.9), where p_{max} and p_{min} are the maximum and minimum number of neurons in the hidden layer. The fix scheme suggests keeping a fixed number of neurons same as the mean of p_{max} and p_{min}. The linear scheme suggests keeping the minimum number of neurons in the first expert and then increase the number of hidden neurons of ε_k by k. Whereas, the random scheme advocates randomly select the number of hidden neurons in $[p_{min}, p_{max}]$. An experimental analysis advocates selecting the hidden neurons using a linear scheme.

$$p^{\varepsilon_k} = \begin{cases} p_{min} + k, & \text{Linear} \\ randi[p_{min}, p_{max}], & \text{Random} \\ \lfloor (p_{min} + p_{max})/2 \rfloor, & \text{Fix} \end{cases} \qquad (7.9)$$

7.3.3 Expert Weight Allocation

In an ensemble-based predictive framework, the choice of weights associated with the individual expert systems significantly impacts the performance. Therefore, an intelligent model should choose the weights carefully. The ELM-based ensemble model finds out the set of suitable weights using one of the nature-inspired algorithms i.e. blackhole optimization algorithm. The population-based optimization algorithm reduces the chances of getting it stuck in local optima as in the case of gradient-based approaches. Moreover, it becomes a better choice if the optimization algorithm is simple and does not use a lot of parameters. Surprisingly, the blackhole optimization algorithm lives up to these factors.

Let $s_i = [\alpha_i^{\varepsilon_1}, \alpha_i^{\varepsilon_2}, \ldots, \alpha_i^{\varepsilon_k}]$ be the ith candidate solution of the pool S which is composed of N such solutions. The term $\alpha_i^{\varepsilon_k}$ represents the weight corresponding to the ε_k. Any solution is randomly generated such that $\alpha_i^{\varepsilon_j} \in [0, 1], \forall j = \{1, 2, \ldots, k\}$ and $\sum_{j=1}^{k} \alpha_i^{\varepsilon_j} = 1$. Let $\hat{x}_t^{s_i} = \sum_{j=1}^{k} \alpha_i^{\varepsilon_j} \times \hat{x}_t^{\varepsilon_j}$ be the weighted forecast value of an ensemble of k networks. The fitness value of s_i can be computed by observing the mean squared error i.e. $f_{s_i} = \frac{1}{m} \sum_{t=1}^{m} (\hat{x}_t^{s_i} - x_t)^2$. As already discussed in previous chapters, the blackhole algorithm selects a solution with least fitness value (because the objective is to minimize the forecast error) i.e. $\mathcal{B} = \min_{i=1,2,\ldots,N} f_{s_i}$ and marks it as a blackhole star which guides the search process further. Further, the algorithm updates the positions of the stars and if any solution enters into the event horizon radius of the blackhole then it gets collapsed and a new solution is generated. The operational summary of the framework is given in Algorithm 7.2.

Algorithm 7.2 Pseudocode of WP$_{eELMNN}$ predictive framework [81]

Input: $X, L, N, T_\varrho, T_{\Xi_{tr}}, T_{\Xi_{ts}}, d, PWS, p_{min}, p_{max}$

Output: $\hat{X}$

1: Select input nodes through autocorrelation analysis
2: Prepare input data according to n
3: **for** each expert (ε_k) **do**
4:    $\omega^{\varepsilon_k} = rand(n, p+1)$
5:    $\tilde{\Omega}^{\varepsilon_k} = \hbar^{\varepsilon_k\dagger}\varsigma$
6: **end for**
7: $S = rand(N, k)$
8: $f_{s_i} = \frac{1}{m}\sum_{t=1}^{m}(\hat{x}_t^{s_i} - x_t)^2$
9: $\mathcal{B} = \min_{i=1,2,...,N} f_{s_i}$
10: **for** each iteration **do**
11:    **for** each star (s_i) **do**
12:       $s_i' = s_i + r_i \times (\mathcal{B} - s_i)$
13:       $f_{s_i^i} = \frac{1}{m}\sum_{t=1}^{m}(\hat{x}_t^{s_i} - x_t)^2$
14:    **end for**
15:    Update Blackhole $\mathcal{B}$
16:    Compute radius of blackhole horizon $\rho = \frac{f_\mathcal{B}}{\sum_{i=1}^{N} f_{s_i}}$
17:    **for** each star (s_i) **do**
18:       $d_{s_i} = f_\mathcal{B} - f_{s_i}$
19:       $s_i = \begin{cases} \Psi & \text{if } d_{s_i} <= \rho \text{ and } s_i \neq \mathcal{B}, \\ s_i & \text{otherwise} \end{cases}$
20:    **end for**
21: **end for**
22: Compute Ξ_{tr}
23: **if** $\Xi_{tr} > T_{\Xi_{tr}}$ **then**
24:    Goto Step 3
25: **end if**
26: Forecast the future workload x_{t+1} and Compute Ξ_{ts}
27: **if** $\Xi_{ts} > T_{\Xi_{ts}}$ **then**
28:    Goto Step 7
29: **end if**

7.4 SHORT TERM FORECAST EVALUATION

The behavior of forecasting approaches on short-term forecasts by measuring the forecast accuracy of the models is evaluated and discussed in this section. The WP$_{ELMNN}$ conducts an experimental analysis to decide the network architecture. The experiments are designed using the D-optimal experiment design method [32]. On the other hand, WP$_{eELMNN}$ creates a simulation environment using $k \in [10, 100], n \in [10, 40], p \in [5, 50], N = 20, G = 100, T_{\Xi_{tr}} = T_{\Xi_{ts}} = 0.007, T_\varrho = 0.1$ parameter settings

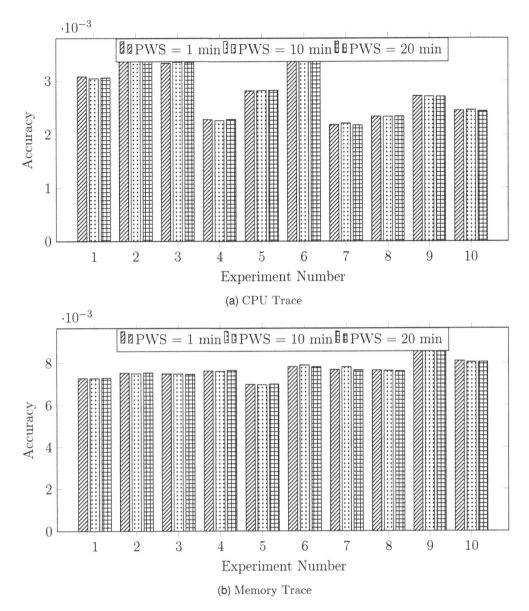

FIGURE 7.6 Network architecture analysis for short term forecast of decomposition predictive framework

and splits the data into 70:30 ratio of training and testing data. It also learns the network architecture through analytical observations.

First, the observations of network architecture optimization are reported. Since WP_{ELMNN} uses 10 different network settings as listed in Table 7.3, the forecast accuracy for all network architectures is observed and analyzed.

The ensemble learning-based predictive framework (WP_{eELMNN}) also learns the network architecture. The number of input neurons is decided based on the data characteristics. If a trace has a higher number of lags with significant autocorrelation,

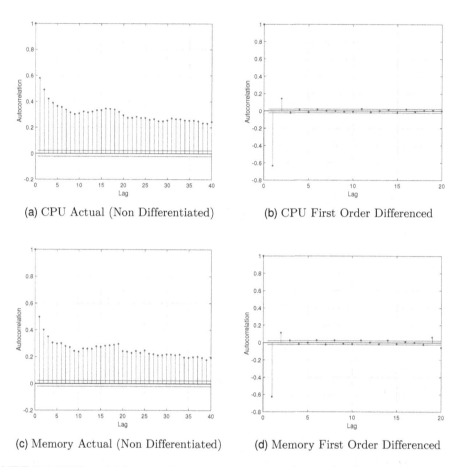

(a) CPU Actual (Non Differentiated)

(b) CPU First Order Differenced

(c) Memory Actual (Non Differentiated)

(d) Memory First Order Differenced

FIGURE 7.7 CPU and Memory data-trace auto-correlation for five minutes prediction interval [81]

the difference operator is applied to reduce the non-linearity from the data. The maximum number of lags to be observed in these experiments is 40. Figure 7.7 shows the autocorrelation in CPU and memory demand traces and their first-order differentiated traces. It can be observed that the autocorrelation in non-differentiated data is present significantly up to 40 lags which is maximum for experiments whereas in differentiated data traces the significant presence of autocorrelation is only for few lags. Since the number of lags with significant autocorrelation defines the number of terms that contribute more in future estimation, the value of n can be decided and set to be the last lag with significant autocorrelation in the data trace.

As discussed in section 7.3.2, the predictive framework also learns the best suitable hidden neuron selection scheme among three heuristics i.e. linear, random, and fixed as shown in eq. (7.9). Figure 7.8 shows the forecast accuracy obtained by applying each hidden node selection scheme on CPU trace. It can be observed that the fix approach produces forecasts with similar errors, and the random selection method generates better forecasts compared to fix. The linear scheme outperforms the other two selection

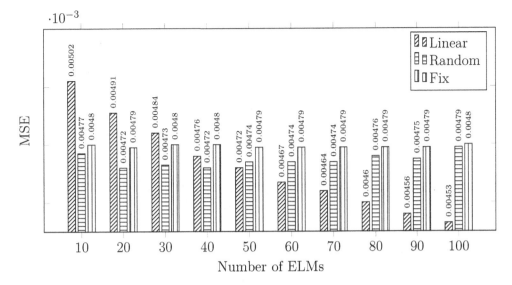

FIGURE 7.8 Short term forecast accuracy of ensemble prediction framework on CPU Trace

strategies when 50 or more expert machines are used to analyze the data due to the fact that the number of hidden neurons gets increases as the number of networks increases. For instance, when $p_{min} = 5, p_{max} = 50$, and $k = 50$ the 50th machine would have $p = 27, p = 30$, and $p = 55$ for fix, random, and linear approaches respectively. Thus, the performance of the approach improves as the number of hidden neurons increases. A similar analysis is conducted for memory trace also and corresponding results are shown in Fig. 7.9. Again, a similar trend is observed and the linear approach gives better estimation after having enough number of experts in the ensemble.

Further, the forecast accuracy of both frameworks on different time intervals is reported in Table 7.4. It can be seen that WP_{ELMNN} predicts the CPU Trace with low forecast errors as compared to WP_{eELMNN}. On the other hand, Memory Trace is better modeled by WP_{eELMNN} as it produced forecasts with lower residuals.

7.5   LONG TERM FORECAST EVALUATION

The network structure optimization results for long-term forecasts are also analyzed for all selected network structures as listed in Table 7.3. Figure 7.10 lists out the forecast accuracy measured in MSE for CPU Trace for all different experiments. It can be observed that a network with $n = 2, p = 10$, and $q = 1$ and 80% training data produces the lowest forecast error for CPU Trace. A trend of increasing the forecast error as the number of input neurons increases on the same size of training data is also detected. The trend is independent to number of neurons in the hidden layer. The forecast accuracy gets improved further if the size of training data is increased. Similarly, a network with $n = 5, p = 10$, and $q = 1$ and 65% training data produces

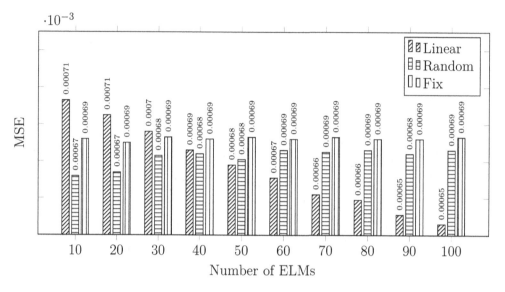

FIGURE 7.9 Short term forecast accuracy of ensemble prediction framework on Memory Trace

the lowest forecast error for Memory Trace. However, the results do not show any trend as opposed to forecasting error on CPU Trace.

TABLE 7.4 Mean squared error of short term forecast of ELM based predictive framework

PWS (min)	WP$_{ELMNN}$		WP$_{eELMNN}$	
	CPU Trace	Memory Trace	CPU Trace	Memory Trace
1	2.175E-03	6.971E-03	0.886E-03	4.640E-04
5	-	-	4.530E-03	6.460E-03
10	2.204E-03	6.957E-03	2.434E-03	1.736E-03
20	2.170E-03	6.983E-03	3.852E-03	2.102E-03

The number of input neurons in WP$_{eELMNN}$ is selected by analyzing the number of lags with significant autocorrelation. It is observed from autocorrelation analysis of workload traces that autocorrelation is present significantly up to 40 lags which is maximum for experiments whereas in differentiated data traces the significant presence of autocorrelation is only for few lags. Since the number of lags with significant autocorrelation defines the number of terms that contribute more in future estimation, the value of n can be decided and set to be the last lag with significant autocorrelation in the data trace.

Similarly, network structure optimization results for WP$_{eELMNN}$ on long-term forecasts are presented. Figure 7.11 shows the forecast accuracy obtained by applying each hidden node selection scheme on CPU trace. It can be observed that the fix approach produces forecasts with similar errors and the random selection method generates better forecasts compared to fix scheme. The linear scheme outperforms the

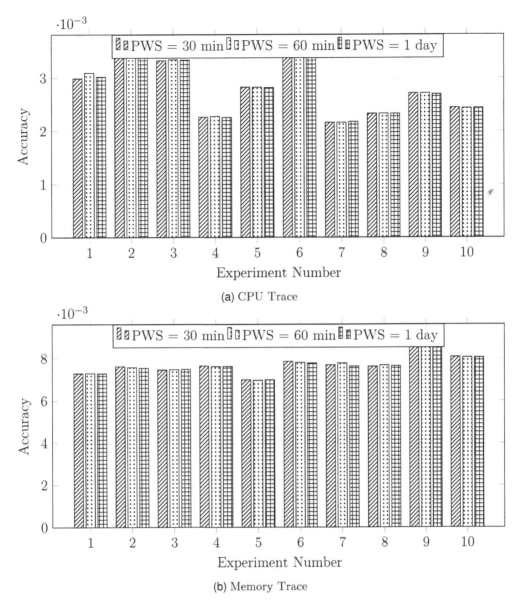

FIGURE 7.10 Network architecture analysis for long term forecast of decomposition predictive framework

other two selection strategies when 50 or more expert machines are used to analyze the data due to the fact that the number of hidden neurons get increases as the number of networks increases. A similar analysis is conducted for memory trace also and corresponding results are shown in Fig. 7.12. Again, a similar trend is observed and the linear approach gives better estimation after having enough number of experts in the ensemble.

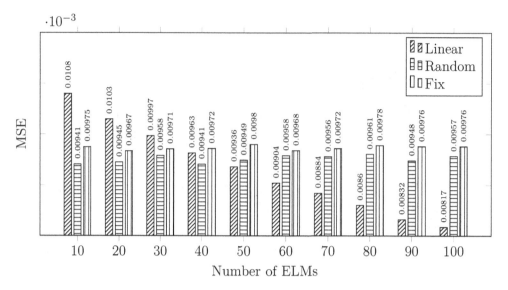

FIGURE 7.11 Long term forecast accuracy of ensemble prediction framework on CPU Trace

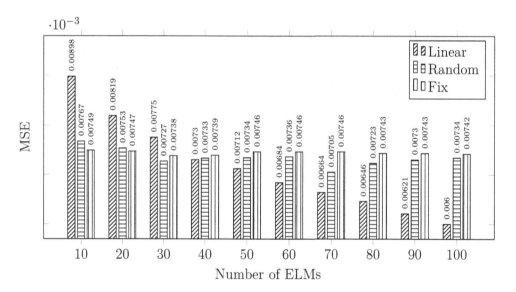

FIGURE 7.12 Long term forecast accuracy of ensemble prediction framework on Memory Trace

Further, the forecast accuracy of both frameworks on different long-term intervals is reported in Table 7.5. It can be seen that WP_{ELMNN} predicts the CPU Trace with low forecast errors as compared to WP_{eELMNN}. On the other hand, Memory Trace is better modeled by WP_{ELMNN} and WP_{eELMNN} as lower residual forecasts are produced by both models for 30 and 60-minute intervals respectively.

TABLE 7.5 Mean squared error of long term forecast of ELM based predictive framework

PWS (min)	WP$_\text{ELMNN}$		WP$_\text{eELMNN}$	
	CPU Trace	Memory Trace	CPU Trace	Memory Trace
30	2.162E-03	6.987E-03	8.930E-03	7.070E-03
60	2.162E-03	6.959E-03	8.170E-03	6.000E-03

7.6 COMPARATIVE ANALYSIS

The performance of both predictive frameworks is compared with the state-of-the-art approaches. Differential evolution, blackhole optimization, support vector regression, ARIMA, linear regression, and backpropagation-based predictive models from the state of art are used for comparison.

Differential Evolution is a numerical optimizer developed by Storn and Price [105]. It encodes the solutions in the form of vectors and manipulates them to explore the search space for a better solution. A large number of variants of the algorithms are available and one such popular variant is used for the purpose of the comparison [74]. The forecast accuracy of the frameworks is also compared with the predictive frameworks that are trained with blackhole neural networks and its variants. Support Vector Regression (SVR) is a supervised learning algorithm that uses an estimation function for training. It creates flexible symmetrical boundaries with a minimum radius around the function and equally penalizes the incorrect estimations. The forecast accuracy of the proposed frameworks is compared with SVR having a parameter grid of cost=100 and gamma=0.0001. In addition, the forecast accuracy on CPU utilization of Google cluster trace is also compared with SVR based prediction model [10]. Linear regression models the relationship among variables using linear prediction models. It has explanatory and dependent variables. A simple linear regression approach is used to compare the forecast accuracy of the predictive frameworks. As opposed to the population-based search algorithm, the backpropagation uses a single solution in the search and finds the better solution by propagating the error feedback which is computed using the concept of gradient [104]. ARIMA is composed of three components viz. Autoregression (AR), Integration (I), and Moving Average (MA). The ARIMA process is a good choice to model non-stationary time series. The ARIMA process integrates the non-stationary data to transform into stationary data by applying the difference operator.

The forecast accuracy on CPU trace is compared using mean squared error and results are listed in Table 7.6. It can be seen that WP$_\text{eELMNN}$ significantly improves the forecast accuracy over WP$_\text{BhNN}$, WP$_\text{BPNN}$, ARIMA, SVR, and LR-based forecasts. It also achieves better forecasts as compared to WP$_\text{SaDE}$ except 1 and 30-minute prediction intervals. However, it is unable to produce better forecasts than WP$_\text{BhNN}^\text{SDL}$ except 60-minute interval forecast. On the other hand, WP$_\text{ELMNN}$ improves the forecast quality as compare to other prediction methods and produces the best forecasts. For instance, it observes a relative reduction in forecast error up to 313.43% and 395.37% against WP_{eELMNN} and WP_{BhNN}^{SDL} respectively.

TABLE 7.6 Forecast accuracy comparison of ELM based predictive models on CPU trace with state-of-art models

PWS (min)	WP_{ELMNN}	WP_{eELMNN}	WP_{BhNN}^{SDL}	WP_{BhNN}	SaDE	BPNN	ARIMA	SVR	LR
1	2.17E-3	8.86E-4	5.14E-4	9.26E-4	8.48E-4	4.41E-3	5.94E-1	1.05E0	1.03E0
10	2.20E-3	2.43E-3	1.80E-3	3.59E-3	2.47E-3	1.42E-2	9.98E-1	7.16E-1	7.18E-1
20	2.17E-3	3.85E-3	3.84E-3	6.28E-3	4.02E-3	2.60E-2	9.96E-1	6.04E-1	6.29E-1
30	2.16E-3	8.17E-3	5.45E-3	1.13E-2	5.49E-3	4.37E-2	8.27E-1	5.34E-1	5.66E-1
60	2.16E-3	8.93E-3	1.07E-2	1.43E-2	1.03E-2	1.03E-1	8.67E-1	6.91E-1	6.89E-1

Similarly, the forecast accuracy of the proposed predictive frameworks is compared on Memory trace and corresponding results are tabulated in Table 7.7. From the results, it can be noticed that both predictive frameworks outperform the prediction schemes based on WP_{BhNN}, WP_{BPNN}, ARIMA, SVR, and LR approaches. It is to note that WP_{ELMNN} outperformed the WP_{SaDE} and WP_{BhNN}^{SDL} on CPU trace but for memory trace, it is unable to outperform these models completely. The short-term forecasts of WP_{SaDE} and WP_{BhNN}^{SDL} on memory trace are observed to be better than WP_{ELMNN} with relatively reduced error up to 96.86%. However, WP_{ELMNN} improves the accuracy of long-term forecasts with a significant margin and relatively reduces the forecast errors up to 69.59%. It should be noted that the memory trace is better modeled by WP_{eELMNN} framework and outperforms the other approaches except few exceptions. These results indicate that the decomposition-based predictive framework is unable to reduce the non-linearities from memory trace as accurately as it reduces from CPU trace. Therefore, it could not outperform the WP_{eELMNN}.

TABLE 7.7 Forecast accuracy comparison of ELM based predictive models on Memory trace with state-of-art models

PWS (min)	WP_{ELMNN}	WP_{eELMNN}	WP_{BhNN}^{SDL}	WP_{BhNN}	SaDE	BPNN	ARIMA	SVR	LR
1	6.971E-3	4.640E-4	2.19E-4	2.48E-4	2.35E-4	1.97E-3	5.06E-1	1.36E0	1.23E0
10	6.957E-3	1.736E-3	3.41E-3	8.56E-3	4.08E-3	2.48E-2	8.89E-1	9.72E-1	9.94E-1
20	6.983E-3	2.102E-3	6.91E-3	1.36E-2	5.54E-3	4.80E-2	5.81E-1	5.18E-1	5.23E-1
30	6.987E-3	7.070E-3	7.08E-3	1.15E-2	9.51E-3	5.16E-2	5.91E-1	5.27E-1	5.30E-1
60	6.959E-3	6.000E-3	1.08E-2	1.09E-2	1.18E-2	8.77E-2	6.10E-1	7.57E-1	7.55E-1

In addition, the forecast accuracy of WP_{eELMNN} is also compared on CPU utilization of Google cluster and Planet Lab traces, and results are listed in Table 7.8, where hyphen (-) indicates that the values are not reported in respective publications. It compares the forecast accuracy with ARIMA and SVR-based prediction schemes using relative mean absolute error on CPU utilization of Google cluster trace and it can be observed that WP_{eELMNN} produces forecasts with better accuracy. The proposed framework observes the reduction in RelMAE up to 19.44% and 19.20%

over ARIMA and SVR-based models. Similarly, a relative reduction up to 99.20% in RMSE is noticed. These promising results convey that the proposed approach produces forecasts with higher accuracy.

TABLE 7.8 Forecast accuracy comparison of ELM based predictive models on Google cluster trace and PlanetLab Trace with state-of-art models

PWS (min)	Google cluster trace(RelMAE)			PlanetLab trace (RMSE)	
	ARIMA [10]	SVR [10]	WP$_{eELMNN}$	Deep Learning [137]	WP$_{eELMNN}$
5	9.72E-01	9.69E-01	7.83E-01	9.17E+00	7.31E-02
30	-	-	8.36E-01	1.03E+01	9.56E-02
60	9.75E-01	-	9.21E-01	9.97E+00	1.03E-01

This chapter discusses two predictive frameworks using extreme learning machines. The WP$_{ELMNN}$ decomposes the traces and trains an ELM for each component. The obtained results achieved an improvement in mean prediction error over state-of-art approaches. The WP$_{eELMNN}$ is composed of multiple base experts i.e. neural networks of different structures trained using the ELM algorithm. Each model analyzes the historical workload information and anticipates the future workload that is further weighted to compute the final forecast. The weights are optimized using a population-based optimization algorithm. The improved forecast quality can be utilized for the low operational cost of the cloud data center by ensuring effective resource management.

CHAPTER 8

Load Balancing

LOAD BALANCING is one of the critical concerns in modern cloud systems. The inefficient resource sharing affects the system performance and introduces various challenges including resource wastage and power consumption in a distributed computing environment. The efficient virtual machine (VM) consolidation is capable of improving cloud performance [12,24,26]. The task of a load balancer is to schedule the computing tasks on the servers such that various parameters including response time, power consumption, resource usage, quality of service are optimized. A load balancing module may optimize one of the various parameters but in practice, it is not always true which means the load balancer tries to optimize multiple parameters. In cloud computing also, a load balancing scheme that can optimize multiple parameters because task scheduling may affect various parameters of the cloud system. In this chapter, we discuss two true multi-objective load balancing models.

8.1 MULTI-OBJECTIVE OPTIMIZATION

A problem which involves multiple criteria to address or optimize can be considered as a multi-objective optimization. It can be denoted as min $\left(f_{cost}^1(v), f_{cost}^2(v), \ldots, f_{cost}^k(v)\right)$, s.t. $v \in V$, where $k > 1$ and depicts the number of objective functions, and V denotes the set of feasible constraint functions or decision vectors. In general, a single solution can not simultaneously optimize each objective of a multi-objective problem. Such problems have a number of solutions commonly referred as nondominated or Pareto optimal solutions. If a solution can not improve any of the objectives without degrading other objective(s), the solution is considered to be nondominated or Pareto optimal.

VM placement is one such problem that has to optimize multiple objectives such as power consumption, SLA violations, resource utilization, etc. Finding an optimal solution to allocate n virtual machines on m servers is an NP-Complete class of problems [36,88]. Figure 8.1 depicts the importance of optimal VM placement. Let a data center contains six servers and it hosts six virtual machines. Let 32%, 58%, 78%, 62%, 14%, and 27% be the resource utilization of each server (Fig. 8.1a). It is to note that the average resource utilization of the data center, in this case, would be 45.16% as all servers are active. Two other possible allocations of these virtual machines are given in Figs. 8.1b and 8.1c and their average resource utilization is 67.75% and 90.33% respectively. More interestingly, these allocations help in saving the electricity consumption as a lower number of servers is active as opposed to the first allocation shown in Fig. 8.1.

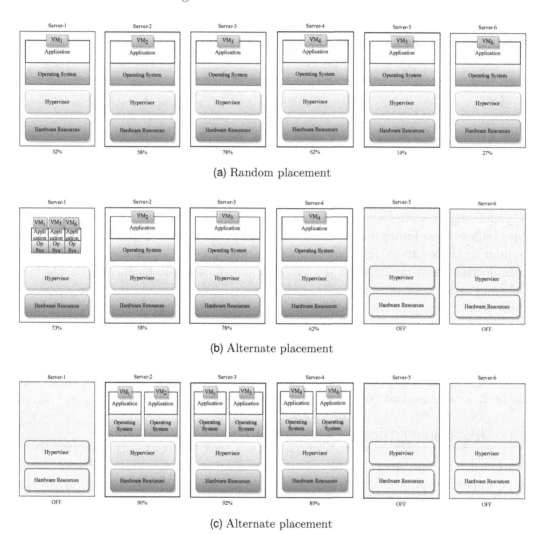

(a) Random placement

(b) Alternate placement

(c) Alternate placement

FIGURE 8.1 An illustration of virtual machine placement scenarios

8.2 RESOURCE-EFFICIENT LOAD BALANCING FRAMEWORK

The priorities of cloud users and service providers are always different. For instance, the users will always expect uninterrupted services at a reasonable cost whereas a service provider will always look for maximizing the number of users and his financial gains. A service provider can increase the number of users only if it can provide the service in accordance with the promised quality of services (QoS) and service level agreements (SLAs). In order to match the user expectations, the service provider has to deploy a large amount of resources. Moreover, it has to ensure the lower operational cost to gain some financial profits. In this section, we will discuss one such framework (RELB) that optimally places the workloads on cloud servers. The framework emphasizes on higher resource utilization and lower power consumption and finds an optimal allocation using a genetic algorithm-based allocation scheme as depicted in Fig. 8.2.

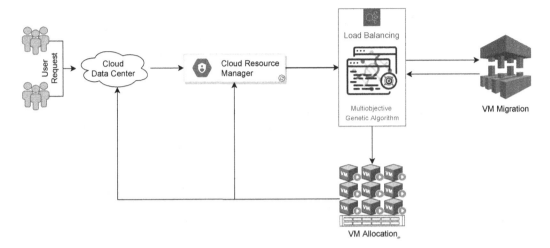

FIGURE 8.2 Resource-efficient load balancing framework design

Let $V_1, V_2, \ldots, V_n$ be the n virtual machines and $U_1, U_2, \ldots, U_p$ be the p cloud users. Considering that a data center DC contains m servers $(S_1, S_2, \ldots, S_m)$ and hosts these n VMs which are owned by p users. If a server (S_j) hosts one or more virtual machines then the server is an active machine i.e. $\beta_j = 1$. Similarly, if a server (S_j) does not host any virtual machine then the server is in the ideal state i.e. $\beta_j = 0$. Let α be the mapping of virtual machine placement such that α_{ij} represents the placement of virtual machine V_i on server S_j i.e. $\alpha_{ij} = 1$ if V_i is hosted on S_j, otherwise $\alpha_{ij} = 0$.

Let S_j^C and S_j^M be the CPU and Memory capacity of S_j. Let V_i^C and V_i^M be the CPU and Memory usage of V_i. The usage of each resource is monitored independently i.e. the CPU and Memory usage is monitored using eqs. (8.1) and (8.2) respectively.

$$\Omega_j^C = \frac{\sum_{i=1}^n \alpha_{ij} \times V_i^C}{S_j^C} \tag{8.1}$$

$$\Omega_j^M = \frac{\sum_{i=1}^n \alpha_{ij} \times V_i^M}{S_j^M} \tag{8.2}$$

This framework monitors resource usage using CPU and Memory utilization. However, the framework is general and resource usage can be monitored using any number of resources. The framework aims to maximize the resource usage of the data center (Ω_{DC}) which is calculated using eq. (8.3), where $|N|$ is the resource count being monitored that is two in this framework.

$$\Omega_{DC} = \frac{\sum_{j=1}^m \Omega_j^C + \sum_{j=1}^m \Omega_j^M}{|N| \times \sum_{j=1}^m \beta_j} \tag{8.3}$$

Furthermore, an ample amount of heat gets generated in the operation of data center of a large scale. Thus, the data center has to maintain the operational temperature for the smooth operation of the data center and it uses a major part of its total consumed electricity for this. After the infrastructure cooling, electricity is consumed by the CPU most [108]. In general, the energy-saving approaches follow the CPU states i.e. they check whether the CPU is busy or ideal. The amount of energy consumption by a busy processor depends on several variables such as the rate of utilization. The resource manager switches off few components of an ideal processor and its operating frequency is reduced. Thus, an ideal processor helps in saving electricity consumption. This framework uses the power consumption modeling that measures the power consumption based on processor utilization [94, 120]. Let P_j^{max} and P_j^{min} be the maximum and minimum power consumption of s_j respectively. Similarly, let P_j^{idle} be the amount of power consumed by s_j in its ideal state. Thus, the amount of power consumption of a server s_j can be calculated using eq. (8.4). Furthermore, the total power consumed by a data center equipped with m servers can be modeled using eq. (8.5) and this framework aims to minimize the consumption of the power of data center (P_{DC}).

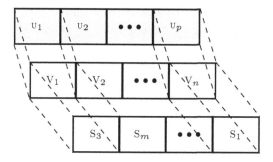

FIGURE 8.3 Chromosome encoding for VM placement

$$P_j = (P_j^{max} - P_j^{min}) \times \Omega_j + P_j^{idle} \tag{8.4}$$

$$P_{DC} = \sum P_j; \quad \forall j = \{1, 2, \ldots, m\} \tag{8.5}$$

$$\sum_{i=1}^{n} v_i^C \times \alpha_{ij} \leq s_j^C; \quad \forall j \in \{1, 2, \ldots, m\} \tag{8.6}$$

$$\sum_{i=1}^{n} v_i^M \times \alpha_{ij} \leq s_j^M; \quad \forall j \in \{1, 2, \ldots, m\} \tag{8.7}$$

The model initialized a set of N random solutions, where each solution places i^{th} virtual machine to randomly selected j^{th} physical machine that satisfies the constraints listed in eqs. (8.6) and (8.7) and turns the server status to active. Figure 8.3 shows

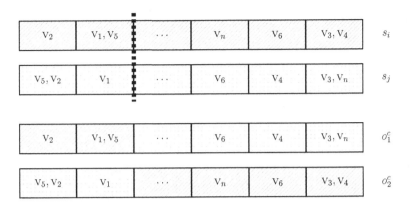

FIGURE 8.4 Single point crossover operator for VM placement

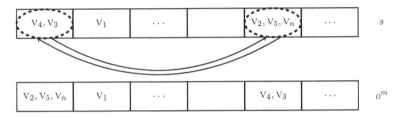

FIGURE 8.5 Swapping based mutation operator for VM placement

the chromosome encoding that allocates each VM on one of the available servers. Each solution (s_i) is evaluated using a cost function (Ψ) that computes the fitness values $(f_{s_i}^{\Omega}$ and $f_{s_i}^{P})$ associated with both objectives. An iterative process that includes the recombination operators such as crossover and mutation are implemented to optimally place the virtual machines on the servers of the data centers.

This framework implements the single-point crossover that switches the tails of both parents as shown in Fig. 8.4. Furthermore, the framework implements the swapping-based mutation operator $(\mathcal{Z})$ that exchanges the two distinct randomly selected VMs allocation as shown in Fig. 8.5. During the course of reproduction, the approach may generate a number of infeasible allocations. Such allocations are migrated as per the availability of the resources to turn the infeasible solutions into feasible solutions by implementing *feasibleAllocation* operation as highlighted in Algorithm 8.1 (line 17). The newly generated offspring solutions (O) are evaluated on the cost function and respective fitness values are observed. The genetic algorithm applies a selection operator to choose N solutions among the current population and offspring solutions. Since each solution has two fitness values, a simple selection scheme becomes infeasible. Therefore, the non-dominated sorting is applied to rank the solutions [33]. The selection operator utilizes the concept of dominance to sort the solutions and rank them based on their non-dominance level as illustrated in Algorithm 8.2. The sorted solutions are arranged in multiple Pareto fronts according to their ranks and the first N solutions are selected to participate in the next iterations. The pseudocode of the proposed framework is shown in Algorithm 8.1.

Algorithm 8.1 Resource efficient load balancing framework operational summary [83]

1: Initialize $g = 0$, g_{max}, N
2: Randomly generate N feasible solutions i.e. $S^g = \{s_1, s_2, \ldots, s_N\}$
3: **for** each $i = (1, 2, \ldots, N)$ **do**
4:    $[f_{s_i}^{\Omega}, f_{s_i}^{P}] = f_{cost}(s_i, \text{DC}^{\kappa})$
5: **end for**
6: **for** $g = \{1, 2, \ldots, g_{max}\}$ **do**
7:    $O = []$
8:    **for** each $i = (1, 2, \ldots, N)$ **do**
9:      $idx = rand(1, N)$ s.t. $i \neq idx$
10:      $cp = rand(1, D)$
11:      $o_1 = [s_i(1:cp), s_{idx}(cp+1:D)]$
12:      $o_2 = [s_{idx}(1:cp), s_i(cp+1:D)]$
13:      $O = [O, \mathcal{Z}(o_1), \mathcal{Z}(o_2)]$
14:    **end for**
15:    **for** each $i = (1, 2, \ldots, 2N)$ **do**
16:      **if** o_i is infeasible allocation **then**
17:        $[o_i, \text{DC}^{\kappa}] = feasibleAllocation(o_i, \text{DC}^{\kappa})$
18:      **end if**
19:      $[f_{o_i}^{\Omega}, f_{o_i}^{P}] = f_{cost}(o_i, \text{DC}^{\kappa})$
20:    **end for**
21:    $S^g = [S^g, O]$
22:    $[S^{g+1}] = \mathcal{S}(S^g)$
23: **end for**

8.3 SECURE AND ENERGY-AWARE LOAD BALANCING FRAMEWORK

One of the ideas behind the success of the cloud paradigm is resource sharing among multiple users for better usability of the resources over time. However, the sharing of resources can be exploited for security breaches through several kinds of attacks. The side-channel attack (SCA) is one of such security threat for cloud users. In this attack, the VMs hosted on a server can be victimized and their sensitive information can be stolen through side-channel events if it also hosts the attacker VM [55]. The users may lose trust in cloud infrastructure and service providers if a series of such attacks occur. Thus, it becomes important to minimize the possibility of such attacks, if not completely avoided. In this section, we will discuss about one such mechanism (SEA-LB) which minimizes the SCA possibility. This approach encourages to host the VMs of a user on the same server as much as possible. But this does not mean that the approach is not allowing resource sharing. The SEALB is capable of reducing the attack possibility on the cost of overhead in terms of resource utilization and power consumption. The conflicting optimization functions (maximum resource utilization, minimum power consumption, and minimum number of shared servers) are optimized using a multi-objective algorithm-based scheme.

Algorithm 8.2 Non-dominated sorting based selection operator [83]

1: **for** each s_i^g **do**
2: $\quad S_i = \emptyset$
3: $\quad n_i = 0$
4: $\quad$ **for** each s_j^g **do**
5: $\quad\quad$ **if** $s_i^g \prec s_j^g$ **then**
6: $\quad\quad\quad S_i = S_i \cup \{s_j^g\}$
7: $\quad\quad$ **else if** $s_j^g \prec s_i^g$ **then**
8: $\quad\quad\quad n_i = n_i + 1$
9: $\quad\quad$ **end if**
10: $\quad\quad$ **if** $n_i == 0$ **then**
11: $\quad\quad\quad s_i^g.r = 1$
12: $\quad\quad\quad F_1 = F_1 \cup \{s_i^g\}$
13: $\quad\quad$ **end if**
14: $\quad$ **end for**
15: **end for**
16: $k = 1$
17: **while** $F_k \neq \emptyset$ **do**
18: $\quad Q = \emptyset$
19: $\quad$ **for** each $s_i^g \in F_k$ **do**
20: $\quad\quad$ **for** each $s_j^g \in S_i$ **do**
21: $\quad\quad\quad n_j = n_j - 1$
22: $\quad\quad\quad$ **if** $n_j == 0$ **then**
23: $\quad\quad\quad\quad s_j^g.r = i + 1$
24: $\quad\quad\quad\quad Q = Q \cup \{s_j^g\}$
25: $\quad\quad\quad$ **end if**
26: $\quad\quad$ **end for**
27: $\quad$ **end for**
28: $\quad k = k + 1$
29: $\quad F_k = Q$
30: **end while**
31: S^{g+1} = Select first N solutions from $F_i \quad \forall i = \{1, 2, \ldots, k\}$
32: **return** S^{g+1}

8.3.1 Side-Channel Attacks

Paul Kocher coined the term 'side-channel attack' when he observed the possibility of accessing the security credentials by applying reverse engineering on power consumption and electromagnetic emission data of a computer system [69]. The SCA primarily collects the information during the target machine performs cryptographic operations. The collected information is reverse engineered to get the sensitive information and then the target machine is attacked. A number of researches are conducted to explore the SCA in shared caches and the side channel behaviors including electromagnetic radiation, time, and power are used to attach the encryption approaches $[23, 31, 49, 87, 127, 133]$. Since a number of users share the computing resources of the same server, the shared distributed architectures are ideal candidates for SCA.

Let $S = \{s_1, s_2, s_3, s_4\}$ be the four servers in a data center, and the resource (CPU and Memory) capacity of each server is $s_j^R = 1.0$ unit $\forall j$. Let $P_j^{max} = 100W$, $P_j^{min} = P_j^{idle} = 20W$ be the maximum, minimum, and idle state power consumption of each server. Furthermore, let this data center hosts $v_1, v_2, v_3, v_4, v_6, v_7$, and v_8 virtual machines which are owned by four users (u_1, u_2, u_3, and u_4). Table 8.1 lists the details of VMs including the owner, resource demands, and the server which hosts the VM, where v_i^k represents the ith VM of user k.

TABLE 8.1 Virtual machine details for illustration

VM Owner	Virtual Machine Id	CPU Demand	Memory Demand	Host Server
u_1	$v_1 \rightarrow v_1^1$	0.15	0.20	s_1
u_1	$v_2 \rightarrow v_2^1$	0.25	0.50	s_4
u_1	$v_3 \rightarrow v_3^1$	0.35	0.40	s_2
u_1	$v_4 \rightarrow v_4^1$	0.20	0.20	s_2
u_2	$v_5 \rightarrow v_1^2$	0.25	0.75	s_1
u_2	$v_6 \rightarrow v_2^2$	0.30	0.60	s_3
u_3	$v_7 \rightarrow v_1^3$	0.50	0.45	s_4
u_4	$v_8 \rightarrow v_1^4$	0.25	0.25	s_2

The data center uses 70% of the resources, consumes $296W$ of electricity, and 75% of servers host the machines of multiple users. In this framework, a machine is referred to as a conflicting server if it hosts the virtual machines of multiple users. The above-mentioned workloads can be placed differently and this framework suggests placing virtual machines as $v_{11}^1, v_{21}^1, v_{33}^1, v_{41}^1, v_{52}^2, v_{64}^2, v_{73}^3, v_{84}^4$. The data center observes the 69.38% resource utilization, $302W$ power consumption, and 50% of servers are conflicting machines. It is interesting to note that the framework reduces the conflicting servers on the cost of a little overhead in terms of power consumption and resource utilization. For instance, the above placement reduces the conflicting servers by 33.33% on the cost of 2.03% higher power consumption and 0.89% lower resource utilization. The above-mentioned illustration is visually depicted in Fig. 8.6, where servers with thicker boundary are the conflicting servers, and shaded servers, are overloaded.

8.3.2 Ternary Objective VM Placement

A *conflicting* server ($\bar{s}_j$) is an ideal prospect for the occurrence of SCA because it hosts the virtual machines owned by multiple users. From the attacker's perspective also, a machine hosting the virtual machines of a large number of users becomes an ideal machine to target due to the fact that compromising such machines would lead to stealing more information. Theoretically, the SCAs can be avoided if machines do not share the resources with different virtual machines which is against the key idea

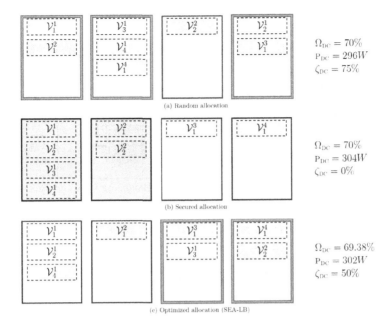

FIGURE 8.6 Three cases of VM allocation

behind cloud technology. Therefore, a balanced approach is required to deal with the attack threats and resource sharing among the users. The SEA-LB targets to minimize the number of conflicting servers while ensuring reasonable resource utilization which means that the framework improves the security without compromising the other aspects much.

Let γ be a matrix that shows the assignment mapping of virtual machines to the servers, i.e. $\gamma_{kj} = 1$ indicates that at least one of the VMs owned by u_k is hosted on s_j and $\gamma_{kj} = 0$ indicates the otherwise case. Using the above representation, $\sum_{k=1}^{p} \gamma_{kj}$ would be the number of distinct users whose virtual machines are hosted on s_j. Thus, the presence (%) of conflicting servers can be calculated using eq. (8.8).

$$\zeta_{\text{DC}} = \frac{\sum_{j=1}^{m} \sum_{k=1}^{p} \gamma_{kj}}{|s|} \times 100 \qquad \forall \sum_{k=1}^{p} \gamma_{kj} > 1 \qquad (8.8)$$

In a large-scale data center, if you want to lower down the number of conflicting servers, the number of active servers will increase. It is to note that the resource utilization and power consumption parameters will be worsen as the number of active servers will increase. Thus, the objectives under consideration are conflicting in nature which means if you improve one objective other will get worse and vice-versa. This framework finds an approximated solution where the virtual machines are placed such that the allocation is secure, energy-aware, and balanced. In order to find such solution, the framework uses one of the most popular and widely used multi-objective algorithms i.e. NSGA-II (Nondominated Sorting Genetic Algorithm-II). Algorithm 8.3 shows detailed pseudocode of the designed framework.

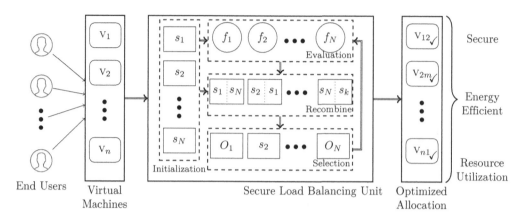

FIGURE 8.7 Secure and energy-aware load balancing framework design [122]

Algorithm 8.3 Secure and energy-aware load balancing framework operational summary [122]

1: Initialize random allocations $(s_1, s_2, \ldots, s_N)$
2: **for** each s_i **do**
3: $\quad [f_{s_i}^{\Omega}, f_{s_i}^{P}, f_{s_i}^{\zeta}] = f_{cost}(s_i, DC^{\kappa})$
4: **end for**
5: **for** each s_i and s_j; $\quad \forall j \in \{1, 2, \ldots, N\}$ and $i \neq j$ **do**
6: $\quad S_i = \emptyset; n_i = 0; S_i = S_i \cup \{s_j\} \quad \forall s_i \prec s_j$
7: $\quad s_i.r = 1; F_1 = F_1 \cup \{s_i\} \quad$ if$(n_i == 0)$
8: **end for**
9: **while** $F_i \neq \emptyset$ **do**
10: $\quad$ **for** $s_i \in F_i$ and $s_j \in S_i$ **do**
11: $\quad\quad n_j = n_j - 1; s_j.r = i + 1; Q = Q \cup \{s_j\} \quad$ if $(n_j == 0)$
12: $\quad$ **end for**
13: **end while**
14: **while** Termination criteria **do**
15: $\quad O_c = \mathcal{Q}(s, DC^{\kappa}); O_m = \mathcal{Z}(s, DC^{\kappa})$
16: $\quad O_c = feasibleAllocation(O_c); O_m = feasibleAllocation(O_m)$
17: $\quad [f_{o_i}^{\Omega}, f_{o_i}^{P}, f_{o_i}^{\zeta}] = f_{cost}(o_i, DC^{\kappa}) \quad i \in \{1, 2, \ldots, l(O_c)\}$
18: $\quad [f_{o_j}^{\Omega}, f_{o_j}^{P}, f_{o_j}^{\zeta}] = f_{cost}(o_j, DC^{\kappa}) \quad j \in \{1, 2, \ldots, l(O_m)\}$
19: $\quad s = \mathcal{S}(s, O_c, O_m); F_1 = \mathcal{NDS}(s)$
20: **end while**

Figure 8.7 shows the system's load balancing unit which shows four major operations involved in the optimization. First, it generates N random solutions which consist of the random allocation of the virtual machines after satisfying the constraints given in eqs. (8.6) and (8.7) corresponding to CPU and Memory resource availability. A feasible VM allocation is represented using s_i s.t. $i = (1, 2, \ldots, N)$ and each cell or gene represents the placement of a virtual machine on a specific machine (s). Then, every solution is assessed on different objective functions i.e. the cost values associated with every objective function are calculated, where $f_{s_i}^{\Omega}, f_{s_i}^{P}$, and $f_{s_i}^{\zeta}$ represent the cost values for resource utilization, power consumption, and security respectively. Further, a dominance level is computed for each solution and they are sorted according to their dominance level. The non dominated solutions are further added into a Pareto front (F_1). The solution s_j is dominated by s_i if s_i is better than s_j on at least one of the objective values and the same or better on other objective functions. The framework applies the recombination operators including crossover ($\mathcal{Q}$) and mutation ($\mathcal{Z}$) to explore the search space for finding a better allocation. The framework employs the single-point crossover operation, whereas the mutation operation randomly swaps two distinct allocations (v_i and v_j, $i \neq j$) as shown in Figs. 8.4 and 8.5 respectively. The solutions obtained after recombination may violate the constraints, therefore, the migration operation is employed to convert an infeasible solution into a feasible solution by migrating the infeasible assignments to the feasible locations. The newly generated solutions are evaluated on the objective functions and the better solutions replace their counterparts in the original set of solutions. Thus, the framework executes a series of iterations and reports a set of approximated solutions and user may select any one of them.

8.4   SIMULATION SETUP

The experimental study on the performance of virtual machine placement approaches was conducted using three types of physical machines in a data center. The CPU capacity is measured in number of processing elements (PE) and million instructions per second (MIPS), whereas the memory is measured in mega bytes (MBs). A physical machine is defined as a tuple (PE, MIPS, Memory (MB), P^{max}, P^{min}, P^{idle}). The three different machines are s_{T_1}(2, 2660, 4096, 135, 93.7, 93.7), s_{T_2}(4, 3067, 8172, 113, 42.3, 42.3), and s_{T_3}(12, 3067, 16384, 222, 58.4, 58.4) [11]. The data center hosts four different types of virtual machines which demand computing resources and each VM is a tuple (PE, MIPS, Memory (MB)). The different configurations of the machines are v_{T_1}(1, 500, 512), v_{T_2}(2, 1000, 1024), v_{T_3}(3, 1500, 2048), and v_{T_4}(4, 2000, 3072), where T_i represents i^{th} type or configuration. Hypothetically, the virtual machines running on a data center may belong to either same category or different categories. If every virtual machine hosted on the data center requests the same amount of resources, it is a homogeneous environment. Similarly, if a virtual machine hosted on the data center requests different amounts of resources, it is a heterogeneous environment.

8.5   HOMOGENEOUS VM PLACEMENT ANALYSIS

In this section, we will discuss the performance of the frameworks in a homogeneous workload environment and it also compares the performance with other existing standard approaches of VM placement. Figure 8.9 depicts the resource utilization results for different numbers of VMs in a data center. The SEA-LB significantly improves the performance as it relatively improves the resource utilization up to 46.64%, 46.64%, and 50.99% over the first fit, best fit, and random heuristics respectively. It is interesting to note that the framework could not improve the performance over the first fit for 300 and 500 virtual machines because SEA-LB sorts the servers according to the resource availability and selects the nearest one to host a virtual machine, whereas the first fit approach selects a server to host according to the resource availability and the index of servers. The first fit scheme improves the resource utilization up to ≈ 13%. Figure 8.8 shows the amount of power consumption. Similarly, Fig. 8.10 shows the presence of conflicting servers in a data center for a different number of virtual machines. In a homogeneous environment, the framework could not reduce the presence of a number of conflicting servers.

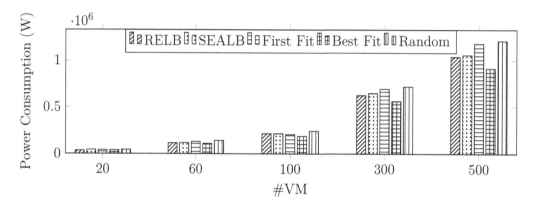

FIGURE 8.8 Power consumption (W) for homogeneous VM requests

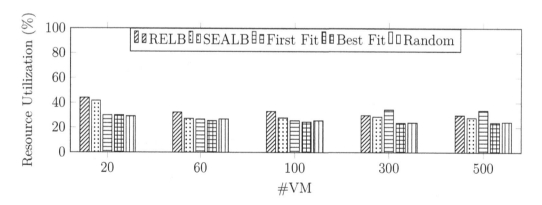

FIGURE 8.9 Resource utilization for homogeneous VM requests

8.6 HETEROGENEOUS VM PLACEMENT ANALYSIS

Similarly, the performance of the approaches is assessed in a heterogeneous environment. The data center is capable of hosting any number of virtual machines up to 500. The resource usage details are shown in Fig. 8.12. The SEA-LB improves the resource utilization up to 119.93%, 84.49%, and 113.96% over first fit, best fit, and random heuristic. It is interesting to note that the RELB outperforms the SEA-LB in terms of resource utilization which indicates the overhead involves in placing the

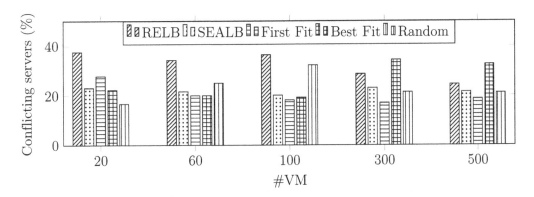

FIGURE 8.10 Presence of conflicting servers (%) for homogeneous VM requests

workloads securely. The SEA-LB is general enough and can handle the situation where the workloads need not be placed securely. In practice, every workload might not need to be placed securely due to several reasons including insensitive data, cost overhead, and various others. Figure 8.11 shows the power consumption of every approach. In this case, the power consumption of the SEA-LB is better than random and best fit placement schemes. Similarly, Fig. 8.13 shows the details of the presence of conflicting servers and it can be noticed that the SEA-LB reduces the chances of SCA by minimizing the number of conflicting servers.

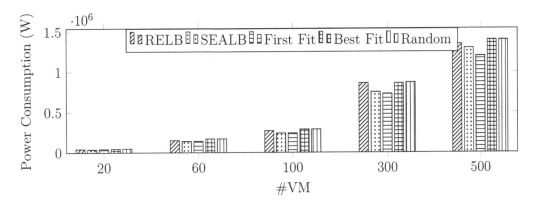

FIGURE 8.11 Power consumption (W) for heterogeneous VM requests

In a cloud environment, the private information of the users is always at risk due to the fact that the resources are shared among other users. In such scenario, a malicious user may initiate the SCA to steal sensitive information. Thus, the cloud service providers have to ensure the security of their user's data. The framework discussed in this chapter is capable of improving security as well as maintaining resource usage and power consumption.

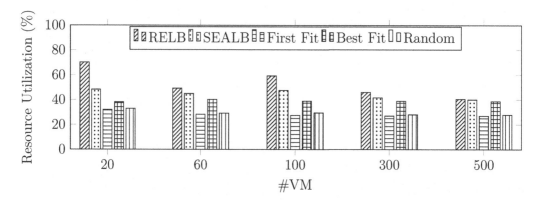

FIGURE 8.12 Resource utilization for heterogeneous VM requests

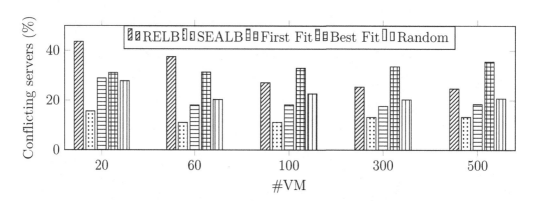

FIGURE 8.13 Presence of conflicting servers (%) for heterogeneous VM requests

CHAPTER 9

Summary

C LOUD COMPUTING is emerged as a solution to various challenges in the computing world including big data management, computing resources availability from
anywhere at anytime, upfront cost management, etc. However, cloud resource management is critical in order to leverage the benefits of cloud paradigm. This book talks
about various recent mechanisms involved in cloud resource management including
workload prediction and load balancing. The time series and neural network-based
forecasting models are explored and their performance on workload prediction is monitored through comparative experimental studies. The forecast accuracy is measured
and compared using statistical tests. Furthermore, the load balancing approaches
developed using multi-objective genetic algorithms are discussed.

It starts with the introduction of cloud computing, followed by the importance
of cloud management. Further, two important aspects involved in cloud resource
management are defined and discussed. The machine learning approaches used to
address the workload prediction and load balancing issues are discussed, followed
by the experimental settings and analysis environments. Time series analysis models
are first discussed and their performance is evaluated on various parameters in
different evaluation environments. The time series analysis has wide applicability and
includes the models that analyze the data in a time domain and extract important
characteristics and meaningful statistics. The models use the extracted information
to estimate the future values based on historic actual values. Afterwards, an error
prevention scheme enabled time series models are discussed and detailed performance
analysis is shown, which shows that the inclusion of error prevention scheme improves
the forecasting ability of the time series analysis model. The error prevention scheme
(EPS) learns the error trend from recent forecasts. The approach analyzes recent
forecasts to measure the average residuals. The observed error trend is referred to as
error preventive score that is leveraged to improve the next forecast. An experimental
study is carried out to analyze the forecast efficacy of error preventive and non-error
preventive time series forecasting models. The autoregressive moving average (ARMA),
autoregressive integrated moving average (ARIMA), and exponential smoothing (ES)
are used to study the effect of the error prevention scheme. The models are examined
for 10, 20, 30, and 60-minute interval forecasts and forecast accuracy is measured
using three error functions i.e. R-Score, sum of elasticity index (SEI), and mean

squared prediction error (MPE). The error preventive models achieved a significant improvement up to 183.9%, 95.4%, 100.0% over R-Score, SEI, and MPE, respectively. In addition, two forecast accuracy metrics, namely predictions in error range (PER) and magnitude of predictions (MoP), are discussed that measure the accuracy of a forecasting model from different dimensions. PER divides the error scale into number of segments and computes the share of forecasts in each segment. The model that produces most of the forecasts with smallest errors is preferred. On the other hand, MoP considers the magnitude of forecasts because it plays a significant role for several applications. For instance, response time is one of the key factors of a cloud system, and a cloud system may prefer higher forecast values to achieve greater user satisfaction. Similarly, a cloud system that prioritizes resource utilization most will not prefer a high forecast. Therefore, the MoP can help in evaluating and selecting the forecasting module as per the application requirements. Further, a statistical analysis using Wilcoxon and Friedman tests is also conducted to validate the observed experimental results.

After discussing the time series models, the metaheuristic algorithms are explored and discussed. The eight different algorithms are discussed that follow the principles from different domains such as swarm behaviour, physics environment etc. These algorithms are being widely used in different applications of optimization. In this book, these algorithms are used to learn the synaptic connection weights of neural networks which form a predictive model for cloud workloads. Their performance is compared at two different levels. First, the performance is compared with the algorithm which belongs to the same category i.e. the algorithm that borrows the principles from the same domain as of the first algorithm. Afterwards, the performance of every algorithm is compared and statistically analyzed with every other algorithm.

The neural networks have been used in various applications such as classification, clustering, prediction and others. A neural network trained with an evolutionary algorithm is also called as evolutionary neural network. Further, the book discusses the predictive models based on evolutionary neural networks. The differential evolution is a simple, reliable and widely used optimization algorithm. First, it extends the self-adaptive differential evolution algorithm that optimizes the mutation strategy from three mutation operators. The optimization process probabilistically selects one of three operators to generate mutant vectors. It records the number of offspring solutions that are successful and unsuccessful in entering into the next iteration generated by each mutation operator. Based on these numbers, the probability of each mutation operator is updated. The approach has used *DE/rand/1, DE/current to best/1,* and *DE/best/1* operators due to the fact that these operators are good for maintaining diversity, convergence property, and optimization problems respectively. Similarly, a second learning algorithm called BiPhase adaptive learning is, discussed that selects the best suitable crossover operator along with mutation operator. Again, three different crossover operators, namely *uniform crossover, heursitic crossover,* and *ring crossover* are used to find the most favourable operator. Both learning algorithms also optimize the mutation and crossover rate. Both algorithms are tested on two benchmark data traces and compared with maximum, average, and backpropagation based network forecasting models. The experiments are conducted for 1, 5, 10, 20, 30, and 60-minute forecast intervals and accuracy is measured using root mean squared error. It was observed that the self-adaptive differential evolution algorithm outperformed the maximum, average, and backpropagation based forecasting approaches and observed the maximum reduction up to 99.85%, whereas the BiPhase adaptive learning algorithm outperformed the self-adaptive algorithm with maximum error reduction up to 80.36% along with other approaches as well.

Furthermore, the predictive models that use a neural network trained using blackhole algorithm are discussed. It also discusses the concept of self-directed learning and its effect on the ability of neural network learning. The model equipped with self-directed learning is referred to as a self-directed forecasting model, whereas a model which learns the network weights without self-directed learning is referred to as a non-directed forecasting model. The non-directed prediction model uses a multilayer perceptron that learns its synaptic connection weights through a population-based optimization algorithm inspired by blackhole phenomenon. Whereas the self-directed learning model uses an improved blackhole algorithm that organizes the population of stars into multiple clusters or subpopulation. The position update procedure of the standard algorithm is also modified by incorporating the local best information to maintain the diversity of the population. The diverse population explores the search space more effectively to find out an approximated global optima. The self-directed learning model also uses the error prevention scheme to improve its forecast accuracy as it enables a model to learn from its past forecasting pattern. Both approaches are tested on six real-world data traces using mean squared error. The forecasts are conducted for 1, 5, 10, 20, 30, and 60-minute forecast intervals. The efficacy of the schemes is validated using a comparative and statistical evaluation. The performance is compared with self-adaptive differential evolution, backpropagation, and deep learning-based prediction schemes and observed a relative improvement of up to 99.99% over respective models.

In decision making, the opinion of multiple experts is always helpful, and ensemble learning is one perfect example of this. This book also discussed two predictive models that learn the network weights using ensemble learning. These models use the extreme learning machines algorithm to learn the network weights. The extreme learning machine commonly referred to as ELM are feed-forward neural networks that learn the synaptic connection weights in a single step. The first model decomposes the workload traces into three components, namely seasonal, trend, and remainder where sum of all three components is the actual workload. All three components are considered as independent trace, and one ELM network is trained for each component. In order to get final forecast, the outcome of each network is aggregated. Next, an ensemble of ELMs is created to achieve better forecast accuracy. The ensemble learning is widely used to improve accuracy. In the proposed model, each expert i.e. ELM network is having different optimized network size i.e. the number of hidden neurons. Each expert learns the network weights and predicts the anticipated workload on the servers. The output of each expert is aggregated using a voting engine that weights the outcome of each network and sums it. The weights for each expert outcome are optimized using a population-based learning algorithm i.e. blackhole algorithm. Both models are evaluated on CPU and memory request traces of Google cluster trace. The proposed models are compared with ARIMA and support vector regression-based forecasting methods. Both proposed approaches find better forecasts than the models based on ARIMA and SVR.

The load balancing is another important view of cloud resource management. In a cloud environment, the load balancing can be applied at different levels including task level, workloads or virtual machine levels, servers level etc. In this book, the two methods dealing with workload balancing by means of optimal placement of workload are discussed. The VM placement is one of the most challenging and complex tasks in the distributed computing networks. Two multi-objective load balancing frameworks for effective cloud resource management are discussed in detail. The first framework emphasizes on efficient and maximum utilization of resources with low power consumption. The model proposes a genetic algorithm-based solution to find an optimal mapping of VMs to physical machines. Since each individual has multiple fitness values, each associated with one objective, the approach ranks the solutions using non-dominated sorting employed in NSGA-II. The framework creates a data centre with multiple servers with different configurations and simulates the approach with homogeneous and heterogeneous virtual machine requests. Next, a multi-objective framework that also addresses the security issue of side channel attacks is discussed. The side channel attack occurs within the virtual machines sharing the same physical resources. Therefore, the framework models the VM placement in such a way that the sharing of physical machines between the number of users is reduced while maintaining resource utilization. It also attempts to lower down the power consumption of the data centre by minimizing the number of active physical machines. The approach is tested with heterogeneous VM requests and found to be better than resource-efficient load balancing framework and other heuristics based load balancing frameworks.

The process automation is one of the key factors in cloud management. The resource provisioning and workload assignment play an important role in the performance of a cloud system. The performance of both processes directly impacts the overall performance of a cloud system. In this book, we have discussed various machine learning approaches to automate both processes with the aim of reasonable accuracy and performance. In the current scenario of the cloud paradigm, where systems are becoming more complex and dynamic, further investigations and developments of the methods are required to address the modern challenges including availability, security, privacy, and integration of cloud systems with other computing paradigms such as fog computing and edge computing.

Bibliography

[1] Traces available in the internet traffic archive. `ftp://ita.ee.lbl.gov/html/`. [Online: accessed on 01-05-2019].

[2] Oludare Isaac Abiodun, Aman Jantan, Abiodun Esther Omolara, Kemi Victoria Dada, Abubakar Malah Umar, Okafor Uchenwa Linus, Humaira Arshad, Abdullahi Aminu Kazaure, Usman Gana, and Muhammad Ubale Kiru. Comprehensive review of artificial neural network applications to pattern recognition. *IEEE Access*, 7:158820−158846, 2019.

[3] Joseph Ackora-Prah, Samuel Asante Gyamerah, and Perpetual Saah Andam. A heuristic crossover for portfolio selection. 8(65):3215−3227, 2014.

[4] Fares Alharbi, Yu-Chu Tian, Maolin Tang, Wei-Zhe Zhang, Chen Peng, and Minrui Fei. An ant colony system for energy-efficient dynamic virtual machine placement in data centers. *Expert Systems with Applications*, 120:228−238, 2019.

[5] Osama Moh'd Alia and Rajeswari Mandava. The variants of the harmony search algorithm: an overview. *Artificial Intelligence Review*, 36(1):49−68, Jun 2011.

[6] Maryam Amiri and Leyli Mohammad-Khanli. Survey on prediction models of applications for resources provisioning in cloud. *Journal of Network and Computer Applications*, 82:93−113, 2017.

[7] E. Assareh, M.A. Behrang, M.R. Assari, and A. Ghanbarzadeh. Application of PSO (particle swarm optimization) and GA (genetic algorithm) techniques on demand estimation of oil in iran. *Energy*, 35(12):5223−5229, 2010.

[8] Rasoul Azizipanah-Abarghooee, Taher Niknam, Farhad Bavafa, and Mohsen Zare. Short-term scheduling of thermal power systems using hybrid gradient based modified teaching-learning optimizer with black hole algorithm. *Electric Power Systems Research*, 108:16−34, 2014.

[9] Indu Bala and Anupam Yadav. Gravitational search algorithm: A state-of-the-art review. In Neha Yadav, Anupam Yadav, Jagdish Chand Bansal, Kusum Deep, and Joong Hoon Kim, editors, *Harmony Search and Nature Inspired Optimization Algorithms*, pages 27−37, Singapore, 2019. Springer Singapore.

[10] F. Baldan, S. Ramirez-Gallego, C. Bergmeir, F. Herrera, and J. M. Benitez. A forecasting methodology for workload forecasting in cloud systems. *IEEE Transactions on Cloud Computing*, 6(04):929–941, oct 2018.

[11] Anton Beloglazov and Rajkumar Buyya. Optimal online deterministic algorithms and adaptive heuristics for energy and performance efficient dynamic consolidation of virtual machines in cloud data centers. *Concurrency Computing : Pract. Exper.*, 24(13):1397–1420, Sep 2012.

[12] P. D. Bharathi, P. Prakash, and M. V. K. Kiran. Virtual machine placement strategies in cloud computing. In *2017 Innovations in Power and Advanced Computing Technologies (i-PACT)*, pages 1–7, April 2017.

[13] Mamun Bin Ibne REAZ and Mohd Marufuzzaman. Pattern matching and reinforcement learning to predict the user next action of smart home device usage. *Acta Technica Corviniesis - Bulletin of Engineering*, 6(3):37–40, 2013.

[14] Ilhem Boussaïd, Julien Lepagnot, and Patrick Siarry. A survey on optimization metaheuristics. *Information Sciences*, 237:82–117, 2013.

[15] George E. P. Box, Gwilym M. Jenkins, and Gregory C. Reinsel. *Time Series Analysis*. John Wiley & Sons, Inc., 2008.

[16] Wayne F. Boyer and Gurdeep S. Hura. Non-evolutionary algorithm for scheduling dependent tasks in distributed heterogeneous computing environments. *Journal of Parallel and Distributed Computing*, 65(9):1035–1046, Sep 2005.

[17] Rajkumar Buyya, James Broberg, and Andrzej M. Goscinski. *Cloud Computing Principles and Paradigms*. Wiley Publishing, 2011.

[18] Rajkumar Buyya, Satish Narayana Srirama, Giuliano Casale, Rodrigo Calheiros, and et al. A manifesto for future generation cloud computing: Research directions for the next decade. *ACM Comput. Surv.*, 51(5):105:1–105:38, Nov 2018.

[19] Rajkumar Buyya, Chee Shin Yeo, Srikumar Venugopal, James Broberg, and Ivona Brandic. Cloud computing and emerging IT platforms: Vision, hype, and reality for delivering computing as the 5th utility. *Future Generation Computer Systems*, 25:599–616, 6 2009.

[20] H. R. Cai, C. Y. Chung, and K. P. Wong. Application of differential evolution algorithm for transient stability constrained optimal power flow. *IEEE Transactions on Power Systems*, 23(2):719–728, May 2008.

[21] K. Chandrasekaran. *Essentials of Cloud Computing*. Chapman and Hall/CRC, 1 edition, 2014.

[22] Mingzhe Chen, Ursula Challita, Walid Saad, Changchuan Yin, and Mérouane Debbah. Artificial neural networks-based machine learning for wireless networks: A tutorial. *IEEE Communications Surveys Tutorials*, 21(4):3039–3071, 2019.

[23] Z. Chen, A. Sinha, and P. Schaumont. Using virtual secure circuit to protect embedded software from side-channel attacks. *IEEE Transactions on Computers*, 62(1):124–136, Jan 2013.

[24] Sakshi Chhabra and Ashutosh Kumar Singh. Dynamic data leakage detection model based approach for mapreduce computational security in cloud. In *Proceedings on 5th International Conference on Eco-Friendly Computing and Communication Systems, ICECCS 2016*, pages 13–19, 4 2017.

[25] Sakshi Chhabra and Ashutosh Kumar Singh. Optimal VM placement model for load balancing in cloud data centers. In *2019 7th International Conference on Smart Computing and Communications, ICSCC 2019*, 6 2019.

[26] Sakshi Chhabra and Ashutosh Kumar Singh. A secure VM allocation scheme to preserve against co-resident threat. *International Journal of Web Engineering and Technology*, 15:96–115, 2020.

[27] R. B. Cleveland, W. S. Cleveland, J.E. McRae, and I. Terpenning. STL: A seasonal-trend decomposition procedure based on loess. *Journal of Official Statistics*, 6:3–73, 1990.

[28] William Jay Conover and William Jay Conover. *Practical nonparametric statistics*. Wiley New York, 1980.

[29] Credit Suisse. 2018 Data Center Market Drivers: Enablers Boosting Enterprise Cloud Growth. `https://cloudscene.com/news/2017/12/2018-data-center-predictions/`, 2017. [Online; accessed 19-05-2019].

[30] M. E. Crovella and A. Bestavros. Self-similarity in world wide web traffic: evidence and possible causes. *IEEE/ACM Transactions on Networking*, 5(6): 835–846, Dec 1997.

[31] A. Cui, Y. Luo, and C. Chang. Static and dynamic obfuscations of scan data against scan-based side-channel attacks. *IEEE Transactions on Information Forensics and Security*, 12(2):363–376, Feb 2017.

[32] P.F. de Aguiar, B. Bourguignon, M.S. Khots, D.L. Massart, and R. Phan-Than-Luu. D-optimal designs. *Chemometrics and Intelligent Laboratory Systems*, 30(2):199–210, 1995.

[33] K. Deb, A. Pratap, S. Agarwal, and T. Meyarivan. A fast and elitist multi-objective genetic algorithm: NSGA-II. *IEEE Transactions on Evolutionary Computation*, 6(2):182–197, April 2002.

[34] Joaquín Derrac, Salvador García, Daniel Molina, and Francisco Herrera. A practical tutorial on the use of nonparametric statistical tests as a methodology for comparing evolutionary and swarm intelligence algorithms. *Swarm and Evolutionary Computation*, 1(1):3–18, 2011.

[35] Saber M. Elsayed, Ruhul A. Sarker, and Daryl L. Essam. An improved self-adaptive differential evolution algorithm for optimization problems. *IEEE Transactions on Industrial Informatics*, 9:89–99, 2013.

[36] F. Farahnakian, A. Ashraf, T. Pahikkala, P. Liljeberg, J. Plosila, I. Porres, and H. Tenhunen. Using ant colony system to consolidate vms for green cloud computing. *IEEE Transactions on Services Computing*, 8(2):187–198, March 2015.

[37] Araf Farayez, Mamun Bin Ibne Reaz, and Norhana Arsad. SPADE: Activity Prediction in Smart Homes Using Prefix Tree Based Context Generation. *IEEE Access*, 7:5492–5501, 2019.

[38] H. Finner. On a monotonicity problem in step-down multiple test procedures. *Journal of the American Statistical Association*, 88(423):920–923, 1993.

[39] Iztok Fister, Iztok Fister, Xin-She Yang, and Janez Brest. A comprehensive review of firefly algorithms. *Swarm and Evolutionary Computation*, 13:34–46, 2013.

[40] Milton Friedman. The use of ranks to avoid the assumption of normality implicit in the analysis of variance. *Journal of the American Statistical Association*, 32(200):675–701, 1937.

[41] Milton Friedman. A comparison of alternative tests of significance for the problem of m rankings. *The Annals of Mathematical Statistics*, 11(1):86–92, 1940.

[42] Alexander A. Frolov, Dusan Husek, and Pavel Yu. Polyakov. Recurrent-neural-network-based boolean factor analysis and its application to word clustering. *IEEE Transactions on Neural Networks*, 20(7):1073–1086, 2009.

[43] Smrithy G S, Alfredo Cuzzocrea, and Ramadoss Balakrishnan. Detecting Insider Malicious Activities in Cloud Collaboration Systems. *Fundamenta Informaticae*, 161(3):299–316, Jul 2018.

[44] Salvador García, Daniel Molina, Manuel Lozano, and Francisco Herrera. A study on the use of non-parametric tests for analyzing the evolutionary algorithms' behaviour: a case study on the CEC'2005 special session on real parameter optimization. *Journal of Heuristics*, 15(6):617, May 2008.

[45] Zong Woo Geem, Joong Hoon Kim, and G.V. Loganathan. A new heuristic optimization algorithm: Harmony search. *SIMULATION*, 76(2):60–68, 2001.

[46] Salyean Giri, Abeer Alsadoon, Chandana Withana, Salih Ali, and A. Elchouemic. Prediction of dementia by increasing subspace size in rank forest. In *2018 IEEE 8th Annual Computing and Communication Workshop and Conference (CCWC)*, pages 255–260. IEEE, Jan 2018.

[47] Fred Glover. Future paths for integer programming and links to artificial intelligence. *Computers & Operations Research*, 13(5):533−549, Jan 1986.

[48] Fred. Glover and Manuel. Laguna. *Tabu search*. Kluwer Academic Publishers, 1997.

[49] M. Godfrey and M. Zulkernine. Preventing cache-based side-channel attacks in a cloud environment. *IEEE Transactions on Cloud Computing*, 2(4):395−408, Oct 2014.

[50] David Goldberg. Genetic algorithms in search, optimization and machine learning. *MA: Addison-Wesley Professional*, ISBN 978-0201157673.

[51] Guang-Bin Huang, Qin-Yu Zhu, and Chee-Kheong Siew. Extreme learning machine: a new learning scheme of feedforward neural networks. In *2004 IEEE International Joint Conference on Neural Networks (IEEE Cat. No.04CH37541)*, volume 2, pages 985−990. IEEE.

[52] Ishu Gupta, Rishabh Gupta, Ashutosh Kumar Singh, and Rajkumar Buyya. MLPAM: A machine learning and probabilistic analysis based model for preserving security and privacy in cloud environment. *IEEE Systems Journal*, 2020.

[53] Ishu Gupta and Ashutosh Kumar Singh. A confidentiality preserving data leaker detection model for secure sharing of cloud data using integrated techniques. In *2019 7th International Conference on Smart Computing and Communications, ICSCC 2019*. Institute of Electrical and Electronics Engineers Inc., 6 2019.

[54] Ishu Gupta, Niharika Singh, and Ashutosh Kumar Singh. Layer-based privacy and security architecture for cloud data sharing. *Journal of Communications Software and Systems*, 15:173−185, 6 2019.

[55] Y. Han, J. Chan, T. Alpcan, and C. Leckie. Using virtual machine allocation policies to defend against co-resident attacks in cloud computing. *IEEE Transactions on Dependable and Secure Computing*, 14(1):95−108, Jan 2017.

[56] Ahmad Hassanat and Esra' Alkafaween. On enhancing genetic algorithms using new crossovers. *arXiv preprint arXiv:1801.02335*, 2018.

[57] Abdolreza Hatamlou. Black hole: A new heuristic optimization approach for data clustering. *Information Sciences*, 222:175−184, 2013.

[58] John H Holland. *Adaptation in natural and artificial systems: An introductory analysis with applications to biology, control, and artificial intelligence*. U Michigan Press, Oxford, England, 1975.

[59] Guang-Bin Huang, Qin-Yu Zhu, and Chee-Kheong Siew. Extreme learning machine: Theory and applications. *Neurocomputing*, 70:489−501, 2006.

[60] Rob J Hyndman and Yeasmin Khandakar. Automatic time series forecasting: the forecast package for R. *Journal of Statistical Software*, 26(3):1−22, 2008.

[61] Rob J. Hyndman and Anne B. Koehler. Another look at measures of forecast accuracy. *International Journal of Forecasting*, 22(4):679−688, 2006.

[62] IDC. Cloud IT Infrastructure Revenues Surpassed Traditional IT Infrastructure Revenues for the First Time in the Third Quarter of 2018, According to IDC. https://www.idc.com/getdoc.jsp?containerId=prUS44670519, 2019. [Online; accessed 19-05-2019].

[63] Gartner Inc. Gartner Forecasts Worldwide Public Cloud Revenue to Grow 17.3 Percent in 2019. https://www.gartner.com/en/newsroom/press-releases/2018-09-12-gartner-forecasts-worldwide-public-cloud-revenue-to-grow-17-percent-in-2019, 2018. [Online; accessed 19-05-2019].

[64] S. M. Islam, S. Das, S. Ghosh, S. Roy, and P. N. Suganthan. An adaptive differential evolution algorithm with novel mutation and crossover strategies for global numerical optimization. *IEEE Transactions on Systems, Man, and Cybernetics, Part B (Cybernetics)*, 42(2):482−500, April 2012.

[65] Luke Jebaraj, Chakkaravarthy Venkatesan, Irisappane Soubache, and Charles Christober Asir Rajan. Application of differential evolution algorithm in static and dynamic economic or emission dispatch problem: A review. *Renewable and Sustainable Energy Reviews*, 77:1206−1220, 2017.

[66] Brendan Jennings and Rolf Stadler. Resource management in clouds: Survey and research challenges. *Journal of Network and Systems Management*, 23(3): 567−619, Jul 2015.

[67] Yılmaz KAYA, Murat UYAR, Ramazan TEKĐN. A novel crossover operator for genetic algorithms: Ring crossover. 2011. https://arxiv.org/abs/1105.0355.

[68] J. Kennedy and R. Eberhart. Particle swarm optimization. In *Proceedings of ICNN'95 - International Conference on Neural Networks*, volume 4, pages 1942−1948, Nov 1995.

[69] Paul C. Kocher. Timing Attacks on Implementations of Diffie-Hellman, RSA, DSS, and Other Systems. In Neal Koblitz, editor, *Advances in Cryptology− CRYPTO '96*, pages 104−113, Berlin, Heidelberg, 1996. Springer Berlin Heidelberg.

[70] Shiann-Rong Kuang, Kun-Yi Wu, Bao-Chen Ke, Jia-Huei Yeh, and Hao-Yi Jheng. Efficient architecture and hardware implementation of hybrid fuzzy-kalman filter for workload prediction. *Integration, the VLSI Journal*, 47(4): 408−416, 2014.

[71] Jitendra Kumar, Rimsha Goomer, and Ashutosh Kumar Singh. Long short term memory recurrent neural network (LSTM-RNN) based workload forecasting

model for cloud datacenters. In *Procedia Computer Science*, volume 125, pages 676–682. Elsevier B.V., 2018.

[72] Jitendra Kumar, Deepika Saxena, Ashutosh Kumar Singh, and Anand Mohan. Biphase adaptive learning-based neural network model for cloud datacenter workload forecasting. *Soft Computing*, 24:14593–14610, 10 2020.

[73] Jitendra Kumar and Ashutosh Kumar Singh. Dynamic resource scaling in cloud using neural network and black hole algorithm. In *Proceedings on 5th International Conference on Eco-Friendly Computing and Communication Systems, ICECCS 2016*, pages 63–67. Institute of Electrical and Electronics Engineers Inc., 4 2017.

[74] Jitendra Kumar and Ashutosh Kumar Singh. Workload prediction in cloud using artificial neural network and adaptive differential evolution. *Future Generation Computer Systems*, 81:41–52, 4 2018.

[75] Jitendra Kumar and Ashutosh Kumar Singh. Cloud resource demand prediction using differential evolution based learning. In *2019 7th International Conference on Smart Computing and Communications, ICSCC 2019*. Institute of Electrical and Electronics Engineers Inc., 6 2019.

[76] Jitendra Kumar and Ashutosh Kumar Singh. An efficient machine learning approach for virtual ma-chine resource demand prediction. *International Journal of Advanced Science Technology*, 123:21–30, 2019.

[77] Jitendra Kumar and Ashutosh Kumar Singh. Adaptive learning based prediction framework for cloud datacenter networks' workload anticipation. *Journal of Information Science and Engineering*, 36:981–992, 2020.

[78] Jitendra Kumar and Ashutosh Kumar Singh. Cloud datacenter workload estimation using error preventive time series forecasting models. *Cluster Computing*, 23:1363–1379, 6 2020.

[79] Jitendra Kumar and Ashutosh Kumar Singh. Decomposition based cloud resource demand prediction using extreme learning machines. *Journal of Network and Systems Management*, 28:1775–1793, 10 2020.

[80] Jitendra Kumar and Ashutosh Kumar Singh. Performance assessment of time series forecasting models for cloud datacenter networks' workload prediction. *Wireless Personal Communications*, 116:1949–1969, 2 2021.

[81] Jitendra Kumar, Ashutosh Kumar Singh, and Rajkumar Buyya. Ensemble learning based predictive framework for virtual machine resource request prediction. *Neurocomputing*, 397:20–30, 7 2020.

[82] Jitendra Kumar, Ashutosh Kumar Singh, and Rajkumar Buyya. Self directed learning based workload forecasting model for cloud resource management. *Information Sciences*, 543:345–366, 1 2021.

[83] Jitendra Kumar, Ashutosh Kumar Singh, and Anand Mohan. Resource-efficient load-balancing framework for cloud data center networks. *ETRI Journal*, 43: 53−63, 2 2021.

[84] Mohit Kumar and S. C. Sharma. Dynamic load balancing algorithm to minimize the makespan time and utilize the resources effectively in cloud environment. *International Journal of Computers and Applications*, pages 1−10, Nov 2017.

[85] P. Ravi Kumar, P. Herbert Raj, and P. Jelciana. Exploring Data Security Issues and Solutions in Cloud Computing. *Procedia Computer Science*, 125:691−697, Jan 2018.

[86] J. K. Lenstra and A. H. G. Rinnooy Kan. Complexity of vehicle routing and scheduling problems. *Networks*, 11(2):221−227, 2006.

[87] L. Lerman and O. Markowitch. Efficient profiled attacks on masking schemes. *IEEE Transactions on Information Forensics and Security*, 14(6):1445−1454, June 2019.

[88] Xin Li, Zhuzhong Qian, Sanglu Lu, and Jie Wu. Energy efficient virtual machine placement algorithm with balanced and improved resource utilization in a data center. *Mathematical and Computer Modelling*, 58(5):1222−1235, 2013.

[89] Lingyun Yang, I. Foster, and J. M. Schopf. Homeostatic and tendency-based CPU load predictions. In *Proceedings International Parallel and Distributed Processing Symposium*, pages 1−9. IEEE Comput. Soc, Apr 2003.

[90] Ang Ee Mae, Wee Kuok Kwee, Pang Ying Han, and Lau Siong Hoe. Resource Allocation for Real-time Multimedia Applications in LTE's Two-level Scheduling Framework. *International Journal of Computer Science*, 43(4):1−11, 2016.

[91] Spyros Makridakis, Steven C. Wheelwright, and Rob J. Hyndman. *Forecasting: Methods and Applications*. Wiley, 3 edition, 1 1998.

[92] Valter Rogério Messias, Julio Cezar Estrella, Ricardo Ehlers, Marcos José Santana, Regina Carlucci Santana, and Stephan Reiff-Marganiec. Combining time series prediction models using genetic algorithm to autoscaling web applications hosted in the cloud infrastructure. *Neural Computing and Applications*, 27(8):2383−2406, Nov 2016.

[93] E. Mezura-Montes and Carlos A. Coello Coello. An empirical study about the usefulness of evolution strategies to solve constrained optimization problems. *International Journal of General Systems*, 37(4):443−473, 2008.

[94] L. Minas and B. Ellison. *Energy Efficiency for Information Technology: How to Reduce Power Consumption in Servers and Data Centers*. Intel Press, 2009.

[95] Banaja Mohanty and Sasmita Tripathy. A teaching learning based optimization technique for optimal location and size of DG in distribution network. *Journal of Electrical Systems and Information Technology*, 3(1):33−44, 2016.

[96] Sathyan Munirathinam and B. Ramadoss. Big data predictive analtyics for proactive semiconductor equipment maintenance. In *2014 IEEE International Conference on Big Data (Big Data)*, pages 893−902. IEEE, Oct 2014.

[97] Sathyan Munirathinam and Balakrishnan Ramadoss. Predictive Models for Equipment Fault Detection in the Semiconductor Manufacturing Process. *International Journal of Engineering and Technology*, 8(4):273−285, Apr 2016.

[98] San Murugesan. Cloud computing: A new paradigm in IT that has the power to transform emerging markets. *International Journal on Advances in ICT for Emerging Regions*, 4(2):4−11, Oct 2008.

[99] Nong Nurnie, Mohd Nistah, King Hann Lim, Lenin Gopal, Firas Basim, and Ismail Alnaimi. Coal-Fired Boiler Fault Prediction using Artificial Neural Networks. *International Journal of Electrical and Computer Engineering (IJECE)*, 8(4):2486−2493, 2018.

[100] Eva Patel and Dharmender Singh Kushwaha. Analysis of workloads for cloud infrastructure capacity planning. In Lakhmi C. Jain, Valentina E. Balas, and Prashant Johri, editors, *Data and Communication Networks*, pages 29−42. Springer Singapore, Singapore, 2019.

[101] G Pavai and TV Geetha. A survey on crossover operators. *ACM Computing Surveys (CSUR)*, 49(4):72, 2017.

[102] Satish Penmatsa, Gurdeep S Hura, and Princess Anne. Adaptive Cost Optimization and Fair Resource Allocation in Computational Grid Systems. In *29th International Conference on Computer Applications in Industry and Engineering (CAINE 2016)*, pages 1−6, Denver, Colorado, USA, 2016.

[103] V. P. Plagianakos, D. K. Tasoulis, and M. N. Vrahatis. *A Review of Major Application Areas of Differential Evolution*, pages 197−238. Springer Berlin Heidelberg, Berlin, Heidelberg, 2008.

[104] J. J. Prevost, K. Nagothu, B. Kelley, and M. Jamshidi. Prediction of cloud data center networks loads using stochastic and neural models. In *2011 6th International Conference on System of Systems Engineering*, pages 276−281, June 2011.

[105] Kenneth V Price. Differential evolution: a fast and simple numerical optimizer. In *Fuzzy Information Processing Society, 1996. NAFIPS., 1996 Biennial Conference of the North American*, pages 524−527. IEEE, June 1996.

[106] K.V. Price, R.M. Storn, and J.A. Lampinen. *Differential Evolution: A Practical Approach to Global Optimization*. Natural Computing. Springer London, Limited, 2005.

[107] A. K. Qin and P. N. Suganthan. Self-adaptive differential evolution algorithm for numerical optimization. In *2005 IEEE Congress on Evolutionary Computation*, volume 2, pages 1785−1791, Sep. 2005.

[108] Dang Minh Quan, Federico Mezza, Domenico Sannenli, and Raffaele Giafreda. T-alloc: A practical energy efficient resource allocation algorithm for traditional data centers. *Future Generation Computer Systems*, 28(5):791–800, 2012. Special Section: Energy efficiency in large-scale distributed systems.

[109] P Herbert Raj, P Ravi Kumar, and P Jelciana. Mobile Cloud Computing: A survey on Challenges and Issues. *International Journal of Computer Science and Information Security*, 14(12):165–170, 2016.

[110] R.V. Rao, V.J. Savsani, and D.P. Vakharia. Teaching-learning based optimization: A novel method for constrained mechanical design optimization problems. *Computer-Aided Design*, 43(3):303–315, 2011.

[111] Esmat Rashedi, Hossein Nezamabadi-pour, and Saeid Saryazdi. GSA: A gravitational search algorithm. *Information Sciences*, 179(13):2232–2248, 2009.

[112] Charles Reiss, John Wilkes, and Joseph L. Hellerstein. Google cluster-usage traces: format + schema. Technical report, Google Inc., Mountain View, CA, USA, November 2011.

[113] I. Rodríguez-Fdez, A. Canosa, M. Mucientes, and A. Bugarín. STAC: A web platform for the comparison of algorithms using statistical tests. In *2015 IEEE International Conference on Fuzzy Systems (FUZZ-IEEE)*, pages 1–8, Aug 2015.

[114] Deepika Saxena and Ashutosh Kumar Singh. Auto-adaptive learning-based workload forecasting in dynamic cloud environment. *International Journal of Computers and Applications*, 2020.

[115] Deepika Saxena and Ashutosh Kumar Singh. Security embedded dynamic resource allocation model for cloud data centre. *Electronics Letters*, 56:1062–1065, 9 2020.

[116] Deepika Saxena and Ashutosh Kumar Singh. Energy aware resource efficient-(EARE) server consolidation framework for cloud datacenter. In *Lecture Notes in Electrical Engineering*, volume 668, pages 1455–1464. Springer, 2021.

[117] Deepika Saxena and Ashutosh Kumar Singh. A proactive autoscaling and energy-efficient VM allocation framework using online multi-resource neural network for cloud data center. *Neurocomputing*, 426:248–264, 2 2021.

[118] Deepika Saxena, Ashutosh Kumar Singh, and Rajkumar Buyya. OP-MLB: An online VM prediction based multi-objective load balancing framework for resource management at cloud datacenter. *IEEE Transactions on Cloud Computing*, 2021.

[119] Stefano Secci and San Murugesan. Cloud Networks: Enhancing Performance and Resiliency. *Computer*, 47(10):82–85, Oct 2014.

[120] N. K. Sharma and G. R. M. Reddy. Multi-objective energy efficient virtual machines allocation at the cloud data center. *IEEE Transactions on Services Computing*, 12(1):158−171, Jan 2019.

[121] Vartika Sharma, Sizman Kaur, Jitendra Kumar, and Ashutosh Kumar Singh. A fast parkinson's disease prediction technique using PCA and artificial neural network. In *2019 International Conference on Intelligent Computing and Control Systems, ICCS 2019*, pages 1491−1496. Institute of Electrical and Electronics Engineers Inc., 5 2019.

[122] Ashutosh Kumar Singh and Jitendra Kumar. Secure and energy aware load balancing framework for cloud data centre networks. *Electronics Letters*, 55: 540−541, 2019.

[123] Ashutosh Kumar Singh and Deepika Saxena. A cryptography and machine learning based authentication for secure data-sharing in federated cloud services environment. *Journal of Applied Security Research*, 2021.

[124] Ashutosh Kumar Singh, Deepika Saxena, Jitendra Kumar, and Vrinda Gupta. A quantum approach towards the adaptive prediction of cloud workloads. *IEEE Transactions on Parallel and Distributed Systems*, pages 1−1, 2021.

[125] Niharika Singh and Ashutosh Kumar Singh. Data privacy protection mechanisms in cloud. *Data Science and Engineering*, 3:24−39, 3 2018.

[126] Rainer Storn and Kenneth Price. Differential evolution: A simple and efficient heuristic for global optimization over continuous spaces. *J. of Global Optimization*, 11(4):341−359, December 1997.

[127] M. Tang, M. Luo, J. Zhou, Z. Yang, Z. Guo, F. Yan, and L. Liu. Side-channel attacks in a real scenario. *Tsinghua Science and Technology*, 23(5):586−598, Oct 2018.

[128] Ruey S. Tsay. Time series and forecasting: Brief history and future research. *Journal of the American Statistical Association*, 95(450):638−643, 2000.

[129] Manu Vardhan, Shrabani Mallick, Shakti Mishra, and D. S. Kushwaha. A Demand Based Load Balanced Service Replication Model. *Journal on Computing*, 2(4):5−10, 2018.

[130] Frank Wilcoxon. Individual comparisons by ranking methods. *Biometrics Bulletin*, 1(6):80−83, 1945.

[131] Bo K Wong, Thomas A Bodnovich, and Yakup Selvi. Neural network applications in business: A review and analysis of the literature (1988–1995). *Decision Support Systems*, 19(4):301−320, 1997.

[132] Alden H Wright. Genetic algorithms for real parameter optimization. In *Foundations of genetic algorithms*, volume 1, pages 205−218. Elsevier, 1991.

[133] R. Xu, L. Zhu, A. Wang, X. Du, K. R. Choo, G. Zhang, and K. Gai. Side-channel attack on a protected RFID card. *IEEE Access*, 6:58395−58404, 2018.

[134] Preetesh K. Yadav, Sourav Pareek, Saif Shakeel, Jitendra Kumar, and Ashutosh Kumar Singh. Advancements and security issues of IoT cyber physical systems. In *2019 International Conference on Intelligent Computing and Control Systems, ICCS 2019*, pages 940−945. Institute of Electrical and Electronics Engineers Inc., 5 2019.

[135] Xin-She Yang. *Nature-Inspired Metaheuristic Algorithms*. Luniver Press, 2008.

[136] Xin She Yang and Xingshi He. Firefly algorithm: recent advances and applications. *International Journal of Swarm Intelligence*, 1(1):36−50, 2013.

[137] Q. Zhang, L. T. Yang, Z. Yan, Z. Chen, and P. Li. An efficient deep learning model to predict cloud workload for industry informatics. *IEEE Transactions on Industrial Informatics*, 14(7):3170−3178, 2018.

[138] Xinqian Zhang, Tingming Wu, Mingsong Chen, Tongquan Wei, Junlong Zhou, Shiyan Hu, and Rajkumar Buyya. Energy-aware virtual machine allocation for cloud with resource reservation. *Journal of Systems and Software*, 147:147−161, 2019.

Index

Printed in the United States
by Baker & Taylor Publisher Services